W. Hayes

Portraits of rare and curious birds, with their descriptions

From the menagery of Osterly Park, in the county of Middlesex

W. Hayes

Portraits of rare and curious birds, with their descriptions
From the menagery of Osterly Park, in the county of Middlesex

ISBN/EAN: 9783742843524

Manufactured in Europe, USA, Canada, Australia, Japa

Cover: Foto ©berggeist007 / pixelio.de

Manufactured and distributed by brebook publishing software
(www.brebook.com)

W. Hayes

Portraits of rare and curious birds, with their descriptions

PORTRAITS

OF

RARE AND CURIOUS BIRDS,

WITH THEIR

DESCRIPTIONS,

FROM THE

MENAGERY OF OSTERLY PARK,

IN THE COUNTY OF MIDDLESEX.

BY W. HAYES, AND FAMILY.

———

LONDON:

PRINTED BY W. BULMER AND CO.
Shakspeare Printing Office:
AND PUBLISHED FOR THE AUTHOR BY R. FAULDER,
NEW BOND-STREET.
1794.

THOMAS PENNANT, ESQ.

OF

DOWNING IN FLINTSHIRE, LLD.

F.R.S. F.S. Nat. Hist. and Antiq. Edinburgh and Perth;
F. S. Rural Oecon. of Odiham; of the Royal Academies of Sciences
of Stockholm, Upsal, and Lund, in Sweden; and of Drontheim,
in Norway, &c. &c. &c.

———

SIR,

On considering to whom I might venture to refer myself for the patronage of the Work now offered to the Public, it will not be a matter of wonder, that a person so eminent in the science of Zoology, and withal so accessible as Mr. Pennant, should have presently occurred to me.

Without hesitation, I was encouraged to apply to you, Sir; and I was not disappointed: with that candour which seldom fails to accompany superior knowledge, you readily allowed me to prefix your most respectable name to my Publication.

Gratitude and honesty, and that powerful motive, self-interest, will concur in exciting me to use my utmost endeavours, that the Work may not discredit the patronage and support of such persons, as have been pleased to honour my list of subscribers with their names.

I feel myself bound, by every possible tie, to exert my utmost efforts to render this at least equal, if not superior, to any periodical publication of this kind. It has hitherto, Sir, merited your attention: and I shall have little doubt of its meeting with the approbation of my other friends, if it shall be found to be executed in such a manner as to deserve a continuance of your favour and encouragement.

I have the honour to be, with the greatest respect,

Sir, your most obliged and

grateful humble servant,

WILLIAM HAYES.

Southall,
Jan. 1. 1794.

ADVERTISEMENT.

WHEN I first engaged in the Drawings which compose the following Work, I had no intention at the moment of presenting them to the Public in the form of a periodical publication; as they were made under the patronage of the late *Robert Child*, Esq. and honoured with a place in his library, at Upton in Warwickshire. But having so large a family, all under my roof, and dependant on my labours, and having only a precarious income, determinable on my decease, I thought it my indispensable duty to prepare against an event which must happen, and perhaps very shortly, by which they would not only be deprived of protection but support.

On this occasion, I had frequent permission to consult a most respectable character in my neighbourhood, concerning the best means of placing my family in such a situation as might afford a reasonable hope of procuring, by their industry and exertion, a decent subsistence.

Upon examining some specimens of their juvenile performances, that Gentleman was pleased to consider them as convincing proofs of early genius, which was worthy of cultivation. It is to his benevolence towards me and my family, that I am indebted for the suggestion of the present plan.

As the greatest care and attention will be observed throughout this Work to render the Plates a faithful representation of the Subject, all that will be necessary in the descriptive part, will be to point out their respective Characters and Qualities, and the Countries of which they are Natives; and in this respect all necessary information will be obtained from the labours of that elegant and enlightened Character, to whom I have the honour and happiness of being permitted to dedicate this Work, and whose zeal and researches have been productive of such entertainment and improvement to the lovers of Natural History.

The First Number is now sent forth into the world; it is the joint exertion of seven of my pupils; and it is their first appeal to the Candour of a generous Public. From that Candour they have every thing to hope, and to them they cheerfully and most respectfully submit their labours; convinced that the motive will plead where the execution may have failed. But if in this first effort of their talents, there should appear a dawn of merit, in some degree deserving the protection of the numerous, respectable, and distinguished Characters who have already honoured this Work with their encouragement, it will be my duty to excite to the utmost of my power their persevering endeavours, to render this Work worthy the very flattering support with which it has been already honoured, and expressive of our united duty, respect, and gratitude.

W. HAYES.

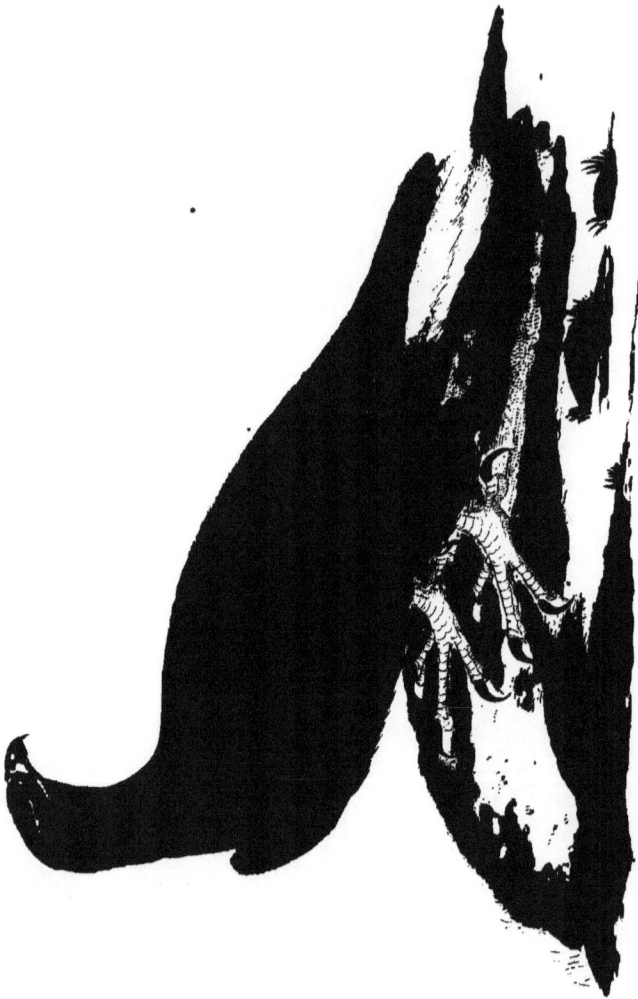

W. & A.M. Meyer

Paris

THE ERNE.

THOUGH some authors have ranked this subject among the Vultures, Mr. Pennant has observed, that it can have no claim to that Genus, as the head and neck are wholly feathered; whereas the head and neck of the Vulture are either quite bare, or only covered with down. The Erne likewise differs from the Eagle, in the want of plumage on the legs, and in the colour of the bill, which in the Eagle is a bluish black, but in the Erne a pale yellow. This bird is found in all the northern parts of Europe, as high as Iceland and Lapmark, in Greenland, Scotland, and the adjacent Isles. It is two feet nine inches long, and almost seven feet from wing to wing: is the first year wholly dusky; in the second, the cinereous colour commences, the tail becomes white, and the end of its feathers, for some time, tipped with black. If not more vigorous than the Eagle, it is at least more bloody and rapacious: it will attack large animals, fish, and birds; especially those which dive, it will watch with great attention, and pounces on them as they rise. It will venture to make its attack on young seals; in this attempt, however, it often suffers by fixing on old ones, which dive into the water, and drown it. The female is larger and more ferocious than the male; they live in pairs, and build their nest in inaccessible rocks. The natives of Greenland use the skin of this bird for cloathing.

THE KING OF THE VULTURES.

Vultur Papa.	*Linn.*
Rex Vulturum.	*Briss.* i. 470.
Roi des Vautours.	*Buff. Ois.* i. 169.

The Count de Buffon, notwithstanding the most diligent inquiry, was not able to discover the least indication of this species among the birds of Asia and Africa; but it being very common in Mexico, New Spain, and the West India Islands, he concludes it to be peculiar to the southern region of the New Continent, though not to be met with in the Old. In size it rather exceeds the hen turkey, measuring from bill to tail two feet three inches. The head is bare: the neck furnished with a tippet of ash-coloured downy feathers, with which it can cover the greatest part of the head when at roost. Its general position is rather stooping than erect; it is extremely sensible of cold, so as to be affected on the first appearance of frost; and notwithstanding the greatest care and attention, it seldom survives the winter of our climate. It is greedy, sullen, deceitful, and rapacious: will prey on fish, lizards, and even on putrid carcases; its sense of smelling is exquisite, and corruption to them hath powerful attractions. In some birds of this species, the body is of a reddish white, in others it is more inclinable to buff. The bill also varies in colour, being in some wholly red, in others of a bright orange at the extremity, and black in the middle: the feet and nails also differ, being of straw colour, with the nails black: in others the feet and nails are of a dull red: the nails in all are short, and very slightly curved. Although this is a most beautiful bird, it is neither elegant, noble, nor generous; and from the nature of its food, it contracts so disgusting a smell, that the Indians, who eat every thing without distinction (even the rattle-snake), will not touch this bird. The most Honourable the Marquis of Stafford did me the honour of presenting me with a pair of these birds, for which I embrace this opportunity of expressing my grateful acknowledgments.

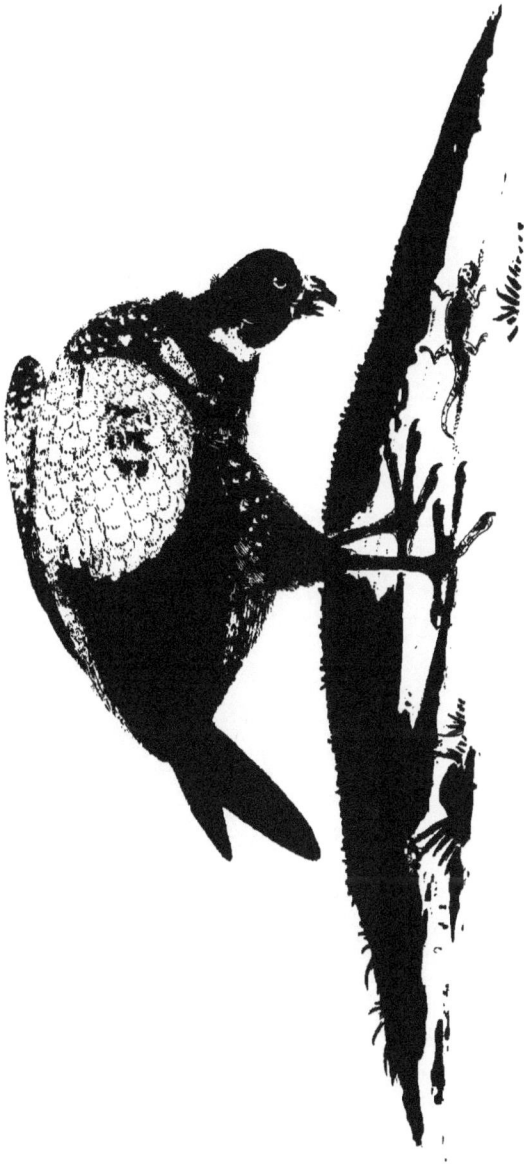

Crowned African Crane

THE CROWNED AFRICAN CRANE.

ARDEA PAVONINA.	*Lin. Syst.* i. *p.* 233. 1.
L'OISEAU ROYAL.	*Buff. Orn.* v. *p.* 511.
CROWNED HERON.	*Lath.* iii. *p.* 34.
CROWNED AFRICAN CRANE.	*Edw.* iv. *p.* 192.

This bird was first brought into Europe by the Portuguese in the fifteenth century; it is a native of Africa, particularly on the Coast of Guinea, Gambia, the Gold Coast, and as far as Cape Verd; where it is so far domesticated as frequently to associate and feed with the common poultry.

I have had frequent opportunities of examining several of these birds in this very curious collection: this beautiful subject was presented to Lady Ducie by the Countess of Chatham, in the most perfect state of plumage.

Though less graceful than the Numidian Crane, it is yet more gentle and familiar. It is much delighted with being taken notice of, and was a constant attendant on those who visited this delightful spot, making the tour of the menagery; with slow but measured steps; and always parting with the company with much apparent regret, which it expressed by raising the head, extending the neck, and making a hoarse unpleasant cry, somewhat resembling the Crane.

When erect, it is near four feet high, and measures from wing to wing five feet four inches. It runs with great rapidity; and not only flies very well, but is able, like the Crane, to sustain it for a long time together. In a state of nature, it frequents the banks of large rivers, and, like the Heron tribe, feeds on small fish, worms, and seeds.

As it had the advantage of every accommodation, it bore the severity of our climate very well. This was the female; but the drawing being taken when the bird was in its most perfect state, it is very little, if at all, inferior to the male, either in size or beauty.

THE NUMIDIAN CRANE.

ARDEA VIRGO.   *Lin. Syst.* i. *p.* 234. 2.

LA GRUE DE NUMIDIE, *vulgairement,*

DEMOISELLE DE NUMIDIE.   *Bris. Orn.* v. *p.* 288.

THIS is the most pleasing bird in the Osterly Collection, and has received the name of Demoiselle, on account of its elegant form, its graceful attitudes, and affected gestures.

The bill is not so long as the Crane, the head (except the crown, which is pale ash colour) and neck are black, the fore part of the neck is adorned with long, soft, narrow-pointed black feathers, which fall over the breast, the rest of the body is of a most delicate bluish ash colour, except the tips of the greater quills and tail, which are dusky; from behind each eye there springs a tuft of long, soft white feathers, of the most delicate texture, which descend in a graceful manner, and which float with the least motion of the wind.

It is gentle and social, apparently much pleased at being admired, and embracing every opportunity of shewing and setting itself off to the greatest advantage to those who seem attracted by its beauty; it accompanies the visiters in their walk in the most graceful manner imaginable, and puts itself into a variety of attitudes, as if it were preparing to entertain the company with a dance.

It is a native of the tropical parts of Africa, on the coasts of Guinea and Tripoli, along the coast of the Mediterranean, and likewise of Egypt. It has been in a manner naturalized in this country. The subject of this Plate, with several others, were hatched and reared in the Osterly Menagery.

Numidian Crane?

Painted . Pheasant .. Male .

THE PAINTED PHEASANT, MALE.

Phasianus Pictus.	*Lin. Syst.* i. *p.* 272. 5.
Le Faisan doré de la Chine.	*Bris. Orn.* i. *p.* 271.
Le Tricolor Huppe de la Chine.	*Bris. Ois.* ii. *p.* 355.

This subject is very justly ranked in the number of the most beautiful birds preserved in this menagery, and in the collections of the curious. It is a native of China, where it is called Kin-ki. In size it is less than the common Pheasant (though more elegantly shaped), being two feet nine inches from the point of the bill to the end of the tail.

As the greatest attention will be observed in colouring the Plates of this work, in order to give a faithful representation of each subject, I shall not in general have occasion to enter into a particular description of the colours; but here it will be necessary to assist the pencil.

The crest is of a most splendid, burnished gold colour, the feathers appearing like silk, which it can erect or depress at pleasure; the cheeks are a tawny flesh colour, thinly beset with feathers; the back of the neck covered with long loose plumage of a bright orange, square at the ends, and marked with transverse bars of a rich velvet black; these likewise can be erected at pleasure, the same as in the domestic cock. In the season of love, when he is addressing the hen, these feathers form a circle from the hind part of the neck to the bill. The feathers from the bottom of the neck to the back are of a deep bronzed green, rounded at the ends, and marked with a circle of black, and which change their position from side to side, according to the attitude of the bird. The coverts of the tail are stiff, long, narrow, and of a bright crimson, and are divided on each side the tail, as expressed in the Plate.

This species is now become naturalized to our climate; they are hardy birds, and require no other attention in breeding them than what is necessary for the common Pheasant.

THE PAINTED PHEASANT, FEMALE.

INSTEAD of being decorated with the gaudy and splendid tints of the male, her plumage is even inferior to the common hen Pheasant, her general colours being a combination of different shades of brown, rufous, tawny, and dusky white. She is likewise smaller than the male, the tail much shorter, and not arched; the legs have no spurs.

The feathers on the head form a small crest, which is only perceptible when she is agitated. She wants likewise the long stiff coverts of the tail, which so very particularly characterize the cock: yet there have been various instances of its transmutation from the dusky colour into the brilliant lustre of the male; one in particular preserved in the menagery of the Countess of Essex, in the space of six years experienced this transformation, and was not to be distinguished from the cock, but from the colour of the eyes, and the shortness of the tail.

It has been generally understood that this remarkable change of colour, whenever it takes place (it being accidental), usually happens to such hens as are four or five years old, when they are neglected by the cock; but Mr. Latham informs us, that the change of plumage in the female is not confined to the Pheasant alone, but has frequently been observed in domestic poultry, when spurs have also sprouted out on the legs of a hen, she crowed at intervals like a cock, and continued to lay eggs, and bred for some years afterwards. A Pea hen of Lady Tyntes', after having many broods, assumed much of the plumage of the cock, with the fine train feathers of the tail.

Painted Pheasant Female.

6 Augt 1790

TOURACO.

Cuculus Persa.	*Lin. Syst.* 171, 17.
Le Coucou verd hupe' de Guine'e.	*Bris. Orn.* iv. 152.
Le Touraco.	*Buff. Ois.* vi. 300.
Touraco Cuckow.	*Lath.* i. *p.* 2. 545.

THE subject of this Plate, which is now before me, is nearly of the size of a magpie, but its tail being frequently spread, increases its apparent bulk, and makes it seem larger than it is in point of fact. It is a native of Africa, and very justly claims a place among the number of the most beautiful birds of this elegant menagery.

The bill is short and compressed sideways, the upper mandible rather arched, but not overhanging the lower. The nostrils are covered with feathers reflected from the forehead; the gape is wide, and separated as far as under the eye, which is lively and uncommonly brilliant; the irides hazle brown, encircled with scarlet caruncles.

The feathers on the crown form a crest, which the bird can raise or depress at pleasure. The plumage is composed of fine soft feathers, or rather fibres, of a delicate silky texture. The legs and feet are of a deep ash colour; the claws are sharp and strong, the toes two forward and two behind.

Edwards is not certain to what genus this subject belongs, but thinks it approaches the nearest to the Cuckow. The Count de Buffon cannot conceive why our Nomenclators range it with the Cuckow, from the common character of having two toes before and two behind, a property belonging to many other birds. But Brisson and Latham decidedly place it with the Cuckow.

RED-BREASTED LONG-TAILED FINCH.

EMBERIZA PARADISÆA.	*Lin. Syst.* i. *p.* 312. 19.
LA VEUVE.	*Bris. Orn.* iii. *p.* 20.
LA VEUVE A` COLLIER D'OR.	*Buf. Ois.* iv. *p.* 155.
WHIDAH BUNTING.	*Lath.* ii. *p.* 1. 178.

THE Count de Buffon treats of eight species of this family, which are generally known by the name of Widow, being a corruption of the word Whidah, a kingdom of Africa, on the Coast of Guinea, where they are common, as well as at Angola, and have likewise been received from Mozambique, a small island lying on the eastern coast of the same continent. They are not confined to Africa only, for they are met with in Asia, and in the Philippine Islands, in the Indian seas.

The subject of this Plate is represented in its summer garb, at which season it acquires the addition of four feathers, which spring from the rump; the two outermost are nearly thirteen inches in length, broad in the middle, narrow at the end; about the middle of these feathers arises a long thread; the two middle feathers are four inches long, very broad, and terminate by a thread; those feathers are marked with undulated transverse bars, and are of a glossy black.

This bird moults twice a year; its first moult is in the spring, at which time it begins to assume its summer dress, but it is not until June that it has recovered its perfect plumage : its second moulting takes place about the beginning of November, it then loses the four feathers above the tail, and the mourning garb of the widow, and by degrees its plumage becomes a mixture of black, brown, tawny, and white, very much resembling the Brambling. The circle of the eye, the bill, and legs experience no variation.

They are very lively sprightly birds, and have an agreeable note, which is supposed to have induced Edwards to class them with the Finches.

GREAT CROWNED INDIAN PIGEON.

COLUMBA CORONATA.	*Lin. Syst.* i. *p.* 282.
LE FAISAN COURONNÉ DES INDES.	*Bris. Orn.* i. 279.
GREAT CROWNED PIGEON.	*Lath.* ii. *p.* 2. 620.

THIS subject is nearly the size of a hen Turkey, in consequence of which Brisson, not having seen the living bird to form a judgment of its instincts and habits, was induced to rank it as a Pheasant, and it was considered as such until Edwards had an opportunity of receiving from Governor Loton a particular history of its character and manners, without which information he never would have conceived that a bird of this magnitude could belong to the family of pigeons.

The note of this bird is cooing and plaintive, but considerably louder, and more expressive of lowing than cooing, than that of the common pigeon.

As the colours of the plumage are exactly given in the Plate, any prolix description would be superfluous. It may perhaps be necessary to observe, that the head is ornamented with a beautiful crest, and the feathers which compose it vary in their length, those in the front not being more than half an inch long, increasing by degrees until they are in length near five inches, the webs being of a loose texture as expressed in the Plate.

These birds are natives of the Molucca Isles, in the Indian seas, under the line, and found in great plenty in New Guinea, from whence they were taken to the Isle of Banda, where they are called by the natives Bululu, and by the Dutch, Kroon-Vogel.

A pair of these birds were presented to Lady Ducie, and kept for some time in the menagery; and it was by her Ladyship's particular order that this drawing was made, at the time they were in the most perfect plumage.

SHAFT-TAILED WHIDAH.

EMBERZA REGIA.	*Lin. Syst.* i. 313.
LA VEUVE DE LA COTE D'AFRIQUE.	*Bris. Orn.* iii. 129.
LA VEUVE A' QUATRE BRINS.	*Buf. Ois.* iv. 158.
THE SHAFT-TAILED BUNTING.	*Lath.* ii. *p.* 1. 183.

THIS subject is the male, and is a native of Africa; it is more rare than the broad-tailed Whidah, and not so large, being about the size of a linnet. It moults twice, in the same manner as the other, and is here represented in its summer dress.

The bill is red, the head black, the throat and parts round the neck, breast, and lower belly, exhibit a blush or pale red, which becomes deeper as it extends behind the neck, which is spotted with black; the lower part of the thighs, and coverts under the tail are black, the four middle feathers of the tail are near ten inches long, and are simple shafts, being only feathered about two inches at the ends, the legs are flesh colour.

In winter this bird loses the long feathers of the tail, its whole plumage changes to a mixture of brown, tawny, and grey, and it can hardly be distinguished from the linnet.

The female is brown, and is not decorated with the four long feathers of the tail, as the male; she likewise experiences the same moultings, but the change is less perceptible than in the male, from the cause above mentioned. In the coure of this work these birds will be given in their winter dress.

CURASSO. MALE.

CRAX GLOBICERA.	*Lin. Syst.* i. *p.* 270. 4.
LE HOCCO DE CURASSO.	*Bris. Orn.* i. *p.* 300.
CURASSOW BIRD.	*Edw. Glean.* ii. *p.* 295.
GLOBOSE CURASSO.	*Lath.* ii. *p.* ii. *p.* 695.

THE Count de Buffon, under the character of Hocco, has given the several species of this family; and his reason for this arrangement is the multitude of names applied by the different savages, in their jargon, as well as by nomenclators, indiscriminately to birds which have many common characters, though distinguished by trifling variations, occasioned by age, sex, or climate; a circumstance very naturally to be expected from a species which is become domesticated.

The subject of this Plate is about the size of a Turkey. The bill is convex, strong and thick; the base covered with a cere, which on the upper mandible swells into a tubercle, or round hard knob, about the size of a cherry, and of a bright yellow, from which it has acquired the name of Globe Curasso.

Another distinguished character that seems peculiar to this bird, is a most beautiful crest, extending from the bill to the back of the head, which it can erect or depress at pleasure, formed of narrow tapering feathers, of different lengths, some almost three inches long, the points reflected, and bent forwards. The head and upper part of the neck is of a rich velvety black, the rest of the body (excepting the lower belly, and under coverts of the tail, which are white) is black glossed with green, purple, and blue, according as it is viewed in different reflections of light.

CURASSO. FEMALE.

Crax rubra.	*Lin. Syst.* i. *p.* 270. 2.
Le Hocco de Perou.	*Bris. Orn.* i. *p.* 305.
Crested Curasso.	*Lath.* ii. *p.* 2. 693.

This is nearly the size of the male, though it varies much in regard to colour. The bill is shaped like his, except that it wants the yellow protuberance, which is not discernible even in the male until the second year, and then varies much in different subjects, according to their age.

The head and neck is black, though not so velvety as in the male; and, in this subject, the rest of the plumage is rather a rich chesnut brown, which becomes paler at the lower belly, and under the coverts of the tail.

These birds are natives of Mexico and Peru. In their wild state they prefer mountainous and retired situations, and perch on the highest trees. They are so extremely stupid as to be insensible of danger; there having been instances where the sportsman has shot several out of the same flock without driving them from their situation, and reloaded his piece as often as was necessary.

When domesticated they become docile and sociable; and though frequently introduced into the menageries of the curious, they are very soon, notwithstanding the greatest care, injured by the dampness of the ground; so that their toes become mortified, which terminates their existence.

From this pair several were bred and raised in this menagery, and in that number a most beautiful one, which, from the variety and richness of its plumage, was termed the Zebra Curasso: and, with other curious specimens preserved in this superb collection, will be given in the course of this publication.

PENCILLED CHINESE PHEASANT. MALE.

PHASIANUS NYCTHEMERUS.	*Lyn. Syst.* i. *p.* 272. 6.
LE FAISAN NOIR ET BLANC DE LA CHINE.	*Buf. Ois.* ii. *p.* 350.
LE FAISAN BLANC DE LA CHINE.	*Bris. Orn.* i. 227.
PENCILLED PHEASANT.	*Lath.* ii. *p.* 11. 719.
BLACK AND WHITE CHINESE PHEASANT.	*Edw.* ii. *p.* 66.

THIS bird is considerably larger than the painted, and even exceeds in size the common, Pheasant.

The head is covered with long feathers, which form a crest, and fall backwards, they are, as well as the throat, breast, belly, thighs, and coverts under the tail, of a full purplish black.

The hind part of the neck, the back, the coverts of the wings and tail, are white, each feather being marked with black lines, which run parallel to the margin of each feather. The two middle feathers of the tail are white, the other feathers are pencilled with black lines.

The eyes are encircled with a carunculated crimson skin, as in the European Pheasant, but rather broader, rises above each eye, and falls on each side below the under mandible; this spot enlarges, and becomes particularly vivid in the season of love. The legs are red, armed with a strong white spur.

This, like the common Pheasant, is always wild and restless; and though in some degree reclaimed, it is never perfectly domesticated, but on every opportunity discovers a vindictive disposition, furiously attacking, with its bill and spurs, whoever approaches or enters its pen.

PENCILLED CHINESE PHEASANT.
FEMALE.

The female is smaller than the male, and differs from him in colour. The bill is horn colour. She has a small tuft of brown feathers, inclining to dull purple, hanging down behind, forming a crest. The eye is yellow, and surrounded by a red skin, which is not so broad or so splendid as that of the male.

The throat, the breast, the belly, and thighs are pale brown, shaded with rufous brown, and marked with irregular transverse bars of different shades of rufous.

The hind part of the neck, the back, the coverts of the wings and tail, are brown tinged with glossy rufous. The two middle feathers of the tail are the same brown, inclining to rufous; the others dull white tinged with brown, and striped with transverse bars of black. The legs the same as the male, but without the spur.

They are natives of China, are now very common with us, breed in our menageries, and are perfectly inured to our climate.

Buffon supposes the White Pheasant to be a native of cold climates, as that of Tartary, and having migrated into the northern provinces of China, has there found a greater plenty of food, more congenial to its nature, so that it has grown to a large size, and is at length become the Pencilled Pheasant.

WAXEN CHATTERER.

AMPELIS GARRULUS.	*Lin. Syst.* i. *p.* 299.
LA JASEUR DE BOHEME,	
BOMBYCELLA BOHEMICA. }	*Bris. Orn. p.* 333.
SILK-TAIL.	*Ray Syn. Av.* 85.
WAXEN CHATTERER.	*Br. Zool. Arct. Zool.*
	Lath. vol. ii. *p.* 1. 91.

ALTHOUGH this subject has been ranked among the British birds, its native climate has not hitherto been determined; this much is certain, they are not stationary, but make their excursions all over Europe: they are found as high as Drontheim, and appear in great numbers during the winter in all parts of Russia, and are there esteemed good food.

They are not unfrequently seen in France, and in Italy. With us they appear in greatest plenty in the northern parts of this island. They have been killed in Northumberland, and Yorkshire, as well as at Eltham in Kent; and the subject of this Plate, together with the female, was shot at Hanwell in Middlesex, Dec. 1783, by Mr. Westbrook, who most kindly indulged me with the liberty of making this drawing. The female was killed; but the male, being only wounded in the wing soon recovered, and became sociable and lively: it gave the preference to fresh juniper berries, rather than any other food. It was presented to Lady Ducie, and placed in the menagery, where it lived some time.

It is reduced on the Plate, the length being almost eight inches; and as its colours are there exactly represented, all that is necessary to add, is that which distinguishes this from every other bird, viz. the small horny red appendages which terminate the tops of six, seven, and sometimes eight of the lesser quill-feathers, that have the colour and gloss of fine sealing wax.

BALTIMORE ORIOLE.

ORIOLUS BALTIMORE.	*Lin. Syst.* i. *p.* 162.
LE BALTIMORE.	*Bris. Orn.* ii. *p.* 109. *Buf.* 3. *p.* 231.
BALTIMORE.	*Arct. Zool.* P. 2. *p.* 302.
BALTIMORE ORIOLE.	*Lath.* i. *p.* 2. *p.* 432.

THIS subject is reduced on the Plate, the length of the species being seven inches. They inhabit many parts of America, from Carolina to Canada, occupying the northern districts in the summer, and returning southward in the winter.

In some places they are, from the brilliancy of their colour, called Fire-Birds, and Fire-hang-nests, their nest being formed in the shape of a pear, open at the top, with a hole at the side, through which the young receive their food, and discharge their excrements.

This nest is formed of a soft downy matter, mixed with wool, woven and lined with hair, and generally supported by two small shoots, which enter the sides of the nest; and it is commonly suspended to the forked branches of the tulip, poplar, and hiccory tree, to which it is fastened with the filaments of some tough plant; and after being thus placed, is perfectly secure from depredations of every kind.

The bill is of a lead colour; the head, the throat, the neck, and the upper part of the back is black; the greater coverts black tipped with white; the quills black, margined with white; the two middle feathers of the tail are likewise black; the rest of the plumage of a most splendid yellow, heightened with orange.

It receives the name of Baltimore from some resemblance, in the distribution of the colours of its plumage, to the arms of Lord Baltimore, who obtained the grant of Maryland.

Baltimore Oriole.
Male.

ALEXANDRINE PARRAKEET.

PSITTACUS ALEXANDRI.	*Lin. Syst.* i. *p.* 141.
PSITTACA TORQUATA.	*Bris. vol.* iv. *p.* 323.
LA GRANDE PERRUCHE A COLLIER D'UN	
ROUGE VIF.	*Buf. vol.* vi. *p.* 141.
RING PARRAKEET.	*Edw. vol. p.* 292.
ALEXANDRINE PARRAKEET.	*Lath. vol.* i. *p.* 234.

THIS species derives its name from having been first noticed during Alexander's Indian expedition. It is nearly the size of a dove-house, or wild pigeon, and with the tail measures sixteen inches.

The bill is red, much hooked, the upper mandible moveable (as in all the parrot tribe),and covered with a cere, in which the nostrils are placed. the tongue large, blunt, rounded, and fleshy; the feet have four toes, two of which are turned backwards; but one of these can be brought forwards occasionally; it is very flexible, and can perform the part of hands in holding any thing, and carrying it to the mouth, also in the act of climbing, which this bird does with great facility, using the bill to assist the feet.

The parrot species is very numerous, they are chiefly confined to the tropical regions of Asia, Africa, and America; a few are met with as far as North Carolina, and at the Straits of Magellan. They are long lived; live chiefly in pairs, but at times assemble in vast numbers; breed in hollow trees, without constructing any nest; and though they lay but two or three (white) eggs at each brood, yet the vast multitude of parrots in the countries which they inhabit, proves to a certainty that they must breed several times annually.

This species inhabits the southern parts of Asia, the adjacent isles, and Ceylon.

MINOR GRAKLE.

GRACULA RELIGIOSA.   *Lin. Syst.* i. *p.* 164.
LE MAINATE.   *Bris. Orn. v.* ii. *p.* 305.
LE MAINATE DES INDES ORIENTALES.   *Buf. Ois.* iii. *p.* 416.
LESSER MINOR, OR MINO.   *Edw.* i. *p.* 17.
MINOR GRAKLE.   *Lath.* i. *part* 2. *p.* 455.

THIS subject is about the size of a blackbird, the length ten inches. The feathers on each side point into the bill as far as the nostrils; on the top of the head they are short, like cut velvet, except just in the middle to the hind head, where they resemble other birds.

On each side of the head is a membrane, in form of a crescent, commencing beneath each eye, and extending to the hinder part of the neck; this membrane is loose at the edge, and irregular in its breadth, of a bright yellow, which suffers a change of colour according to the different seasons, and the various passions by which these birds are actuated, either by anger, or by pleasure.

The whole plumage is black, but more shining on the upper part of the body; the throat, the wings, and the tail, which are beautifully glossed with blue, green, and purple, as exhibited to the view in different lights. The legs and feet are strong, inclining to an orange colour, the claws light brown.

They are very social and lively birds, have great talents for whistling, singing, and speaking; and excel even the parrot for the distinctness of their pronunciation.

They are found in several parts of the East Indies, in the Isle of Hainan, and in almost every island beyond the Ganges. At Java they are common, and are sold there to the Chinese at the rate of five shillings each, for the purpose of keeping them in cages.

SECRETARY.

FALCO SERPENTARIUS.	*Syst. Nat. col. Gmel.* 256.
SECRETARY VULTURE.	*Lath. Syn.* i. 20. 17.
SAGITTARIUS.	*Vosmaer-Monogr. t.* 8.
SECRETAIRE.	*Sonn. Voy.* 87. *t.* 50.
SLAANGEN-VRAATER.	*Sparm Voy.* 1. 154.

THIS very singular bird, although a native of Africa, has not long been known even at the Cape. They are found in the country, a few leagues from the shore; are taken young from the nest, reared, and much valued by the natives, for the purpose of destroying rats, toads, and serpents: the latter it will strike with its wings until it is disabled; then seizing it by the tail, dashes it with great violence on the ground, which it repeats until the serpent is dispatched; on which account they have obtained the name of Slaangen-Vraater, or Serpent Eater, by the natives.

It is rather more than three feet high, when erect; and being not only a new species, but so ambiguous in its form and habits, it has occasioned much uncertainty as to its class, in regard to what family it belongs.

By the form of the bill, it appears to be rapacious, and has been ranked with the Vulture; yet it makes no use of this weapon, either in attack or defence; and its manner, instead of being sullen and cruel (the characteristic of the Vulture), is social, gentle, and inoffensive. It very soon becomes familiar, expressing a grateful attachment and attention to the person who feeds it.

From the extraordinary length of its legs, it might by those unacquainted with its habits be classed with the Waders; but it is most decidedly rapacious. Its strength and defence is in its legs; and what is very remarkable, and observed in no other bird, is its power of striking forwards, never backwards. Dr. J. R. Foster has mentioned a circumstance, which he says was supposed to be peculiar to this bird—that, should it by any accident break a leg, the bone would never unite again.

THE WAX-BILL.

Loxia Astreld.	*Lin. Syst.* i. *p.* 305
Senegalus Striatus.	*Bris. Orn.* iii. *p.* 210.
Le Senegali Raye'.	*Buff. Ois.* iv. *p.* 101.
Wax-Bill.	*Edw. Gl. pl.* 354. *fig.* 2.
Wax-Bill Grosbeak.	*Lath.* ii. *p.* 1. 152.

This bird derives its name from the colour of the bill being of a bright red, resembling sealing-wax; it is represented on the Plate the size of life. A line of bright red extends from the nostrils to the hind head, in which the eyes are placed. The colour of the upper part of the bird is brown, radiated with transverse bars of dusky brown, very delicate on the head and neck; but which increases in breadth and colour on the back, wings, and tail. It becomes much lighter on the throat, breast, and belly. Under the breast is a broken line of red. The lower belly, and coverts under the tail, are black (in the male). The tail is dusky brown, with transverse lines of a deeper shade, and cuneiform. The legs and feet are light brown.

The red-rumped, and white-rumped Gros-beak, described by Latham, are varieties of this kind.

The Count de Buffon, under the characters of Senegalis and Bengalis, describes this and the Amaduvade; and observes, that we should be much mistaken if we inferred from the above names that they are confined to Bengal and Senegal, as they are spread through the greatest part of Asia and Africa, and the adjacent islands.—That described by Brisson came from Java; that by Edwards, from the East Indies. In Senegal, this and other small birds are caught by showing a few grains of millet under a calabash, or large gourd; which is placed on the ground, and raised by a short prop, to which a string is placed, which the person draws at a proper time, and secures whatever is under the calabash.

THE CRANE.

ARDEA GRUS.	*Lin. Syst.* i. *p.* 234. 4.
GRUS.	*Bris. Orn.* v. *p.* 375. 6.
LA GRUE.	*Buff. Ois.* vii. *p.* 287.
COMMON CRANE.	*Lath.* iii. *p.* 1. 40.

THE Crane, originally a native of the north, visits all the temperate climates; it formerly bred in England, as we find in Willoughby there was a penalty of twenty-pence for destroying an egg of this bird; and they likewise spent the winter here, frequenting the fens of Lincolnshire and Cambridgeshire in great flocks. At present the inhabitants of those counties are unacquainted with them, as of late none have been seen, except a single bird shot a few years since near Cambridge; we therefore suppose these birds to have forsaken our island, though no reason has been given for it, as no diminution in the species has been observed, and Linnæus assures us they are as numerous as ever in Sweden.

It is a large stately bird, measuring upwards of five feet, it has an erect, slender, and elegant form, walking with slow and measured steps.

The top of the head is covered with dusky down, sprinkled with hairs or bristles, the hind head bald and red, with a few hairs. On each side is a broad white line the length of the neck, the fore part of which is a deep slate colour approaching to black, as far as the breast; the greater wing coverts, and those farthest from the body are tipped with dusky brown; the bastard wings and quills black; the rest of the plumage a most delicate ash colour: from the pinion of each wing springs an elegant tuft of loose feathers curled at the ends, which the bird can erect or depress at pleasure, but which in a quiescent state hang over and cover the tail. These feathers were formerly held in high estimation, being set in gold, and worn as ornaments in caps.

Such is its astonishing power of wing, that in migrating they frequently soar so high as not to be visible.—Linnæus asserts, at the height of 3 miles.

AMADUVADE.

Fringilla Amandava.	*Lin. Syst.* i. *p.* 319 10.
Bengalus Punctulatus.	*Bris. Orn.* iii. *p.* 206. 62.
Le Bengale Piquete'.	*Buff. Ois.* iv. *p.* 96.
Amadavad Bird.	*Edw. Gl. pl.* 355.
Amaduvade-Finch.	*Lath. vol.* ii. *p.* 1. 311.

This bird is given the size of life, which is nearly that of the wren. The bill is red, and conic as the Finch's, to which family, and not to the Grosbeak, it belongs. The head and throat are of a dull red; the hind part of the neck, the back, wings, and tail, are of a dusky brown; the lower belly and thighs are likewise brown, but paler than the back; the upper coverts of the tail are red, the breast and under coverts of the tail a bright yellow.

Each feather on the side of the neck, the coverts of the wings, the belly, and the end of the tail, are punctulated with white points.

I have had the opportunity of examining many of these birds in this Menagery, as well as in the superb collection of the late Earl of Sandwich, and have remarked they vary in point of colour.

The female is brown, and wants those white points which decorate the male; she differs likewise in other respects, her throat being white; the neck, breast, and belly a pale yellow.

Edwards considers this as the smallest of granivorous birds yet discovered. Its note is sweet and short, often repeated; and he supposes they might, with proper care, breed in this country as the Canaries: he has frequently observed the cock to drive the hen to nest.

Mr. Tunstall remarks, that they become more spotted in proportion to their age; and that one in particular which seemed powdered with white, when first in his possession, had scarce any white spots about it.

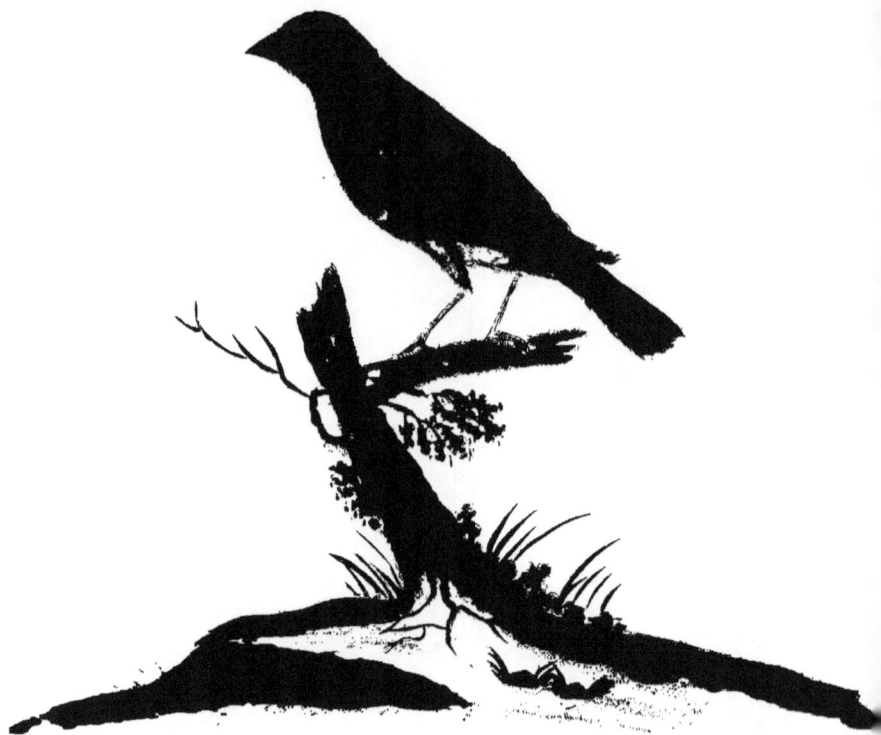

BLACK-FACED BUNTING. MALE.

Embereza Quelea.	*Lin. Syst.* i. *p.* 310.
Le Moineau a bec rouge de Senegal.	*Bris. Orn.* iii. *p.* 110.
Black-Faced Bunting.	*Lath.* ii. *p.* 1. 192.

These birds are represented the size of life. The bill is red, stout, and thick, like that of the Grosbeak, but the formation is truly that of the Bunting; to which family, as Mr. Latham remarks, it certainly belongs.

The forehead is black, the throat and cheeks are likewise black, the back, the coverts of the wings and tail are dusky in the middle, and have a rufous border; the throat, breast, belly, and coverts under the tail are a pleasant blush colour; the quill-feathers dusky brown, with a rufous margin; the tail, consisting of twelve feathers, is likewise dusky brown, with a rufous edge; the legs and feet flesh colour.

A pair of these birds were presented to Lady Ducie, under the appropriated name of Weaver birds; which name they have acquired from this circumstance. A skein of silk having lain accidentally near their cage, and within their reach, they with great art, and indefatigable industry and perseverance interwove it between the wires of their cage, so perfectly as to fill up the side on which they were employed entirely.

BLACK-FACED BUNTING. FEMALE.

THAT the female wants the black about the head and throat, is not the only difference from the male, there are other marks of dissimilitude.

In the male the bill is red, in this it is a pale, yellowish, flesh colour. The head, neck, back, coverts of the wings, and tail, are tawny, each feather dusky in the middle, and margined with a pale tawny colour; the throat, breast, belly, and coverts under the tail, are also of a pale tawny; the quills and tail dusky, with a tawny margin; the legs a yellow flesh colour.

They were naturally docile and familiar birds, and after having lived some time in this menagery, they undoubtedly became more so, by their being accustomed to society, as they never discovered the least uneasiness or apprehension on being frequently examined by the company that frequented this delightful spot.

The hen expressed great fondness and attachment to the cock, always nestling close to him on the perch.

NE'

BLUE-HEADED PARROT.

PSITTACUS MENSTRUUS.	*Lin. Syst.* i. *p.* 148.
LE PARROQUET A TETE BLEUE DE LA GUIANA.	*Bris. Orn.* iv. *p.* 247.
LE PAPEGAI A TETE ET GORGE BLEUE.	*Buff. Ois.* vi. *p.* 213.
BLUE-HEADED PARROT.	*Lath.* i. *p.* 301. *Ed.*
	vii. *p.* 314.

THE Count de Buffon has discriminated this genus into two classes; the first comprehends those of the old world, which he has subdivided into five families; Cockatoos, Parrots, Lories, long-tailed Parrakeets, and short-tailed Parakeets: those of the new world into six families; Maccaws, Amazonians, Cricks, Popinjays, long-tailed Paroquets, and short-tailed Paroquets.

As I shall have an opportunity of introducing a specimen of each family, in the course of this work, I shall then point out their specific characters, in the order they are presented to my Subscribers.

This subject is the size of the dove-house pigeon. The bill is black, with a red spot on both sides of the upper mandible; the eyes are surrounded by a flesh-coloured membrane; the nostrils are placed in a narrow skin at the basis of the upper mandible, pretty close together; on each side the head is a dusky spot, the head and neck are of a fine blue, which inclines to purple on the breast; the wing coverts of a yellow green, the back, belly, thighs, wings, and upper coverts of the tail, of a beautiful green, the under coverts are scarlet, the tail is expanded in order the better to explain the variety and brilliancy of its colours; the legs are grey.

This beautiful Popinjay is a native of Guiana, and though very scarce there, is only valued for the beauty of its plumage, as it cannot be taught to articulate, as most of the parrot tribe do; it is smaller than the Amazon, from which and the Crick it is distinguished by the omission of red on the wings.

YELLOW GOLD-FINCH.

FRINGILLA TRISTIS.	*Lin. Syst.* i. *p.* 320.
LA CHARDONNERET D'AMERIQUE.	*Bris. Orn.* iii. 64.
LA CHARDONNERET JAUNE.	*Buff. Ois.* iv. 212.
GOLDEN FINCH.	*Arct. Zool.* iii. 60.
AMERICAN GOLD-FINCH.	*Catesby* i. 43. *Ed.* 274.
	Lath. ii. 288.

THIS subject is the size of life, and excepting Mr. Pennant, and the Count de Buffon (who in adopting the name of Yellow Gold-finch, very justly characterizes its plumage), all those who have mentioned this bird, call it the American Gold-finch; though this name cannot be strictly proper, unless it can be ascertained that no other Gold-finch exists in that quarter of the world.

They inhabit New York, where they are called York-yellows; they are rare in Carolina, more frequent in Virginia; and Father Charlevoix has met with them in Canada, and other parts of America, they are likewise found at Surinam and Guiana; they are summer birds, and feed on thistles like the Gold-finch.

The bill is of a flesh colour, the irides hazel, the forehead black, the head, neck, back, and breast of a most splendid yellow; the lower belly and coverts under the tail white, the wing coverts black, crossed with bars of white, the edges and tips of the lesser ones white; tail black; legs flesh colour.

Mr. Tunstal, who had several of those birds in his very curious collection, remarks that both male and female constantly lose their yellow in the winter season, and become exactly the colour of the Siskin,* and as constantly recovered their original splendid plumage in the spring.

* Siskin. *Lath.* ii. *p.* 291. 58. B

Yellow Gold Finch.

BLACK-CAPPED LORY.

PSITTACUS DOMICILLA.	*Lin. Syst.* i. *p.* 145.
LE LORY DES INDES ORIENTALES.	*Bris. Orn.* iv. *p.* 222.
LE LORI A` COLLIER.	*Buff. Ois.* vi. *p.* 130.
SECOND BLACK-CAP LORY.	*Edw.* iv. *p.* 171.
PURPLE-CAPPED LORY.	*Lath.* i. *p.* 271.

MR. LATHAM says, this genus of parrots consists of infinite varities, which seem to run so much into each other, as to induce one to think many of them related, though brought from different parts of the world. This assertion, however, must be received with no small hesitation; for as they are considered as articles of traffic by the natives of the tropical regions, they are perpetually carried from one continent to the other for sale.

The name of Lory has been applied to this family in the East Indies, from its cry resembling the word *lory.* They are likewise distinguished from the rest by their plumage, which is chiefly red, of different shades; their bill is also smaller, not so much hooked, but sharper than the parrots.

This species inhabits only the Moluccas, and New Guinea; and if they have been met with elsewhere, as it is known they have not the power of flying to a great distance, it is impossible they should ever have migrated from one continent to the other.

They are very brisk and lively birds, more active than any other parrot; they are gentle and familiar, and are taught, with great facility, to whistle and articulate very distinctly. It is astonishing how soon they can repeat what they hear.

The Count de Buffon supposes that the female wants the yellow crescent on the breast : I rather conjecture his specimen must have been a young bird, before the plumage was perfect, as the bird in Lady Ducie's collection, from which this drawing was made, corresponds exactly with one now in the possession of Lady James, which her Ladyship did me the honour of informing me, had laid an egg since her Ladyship had it in her aviary.

ANGOLA GROSBEAK.

Loxia Angolensis.	*Lin. Syst.* i. *p.* 303.
Black Grosbeak.	*Edw.* vii. *p.* 352.
Angola Grosbeak.	*Lath.* ii. *p.* 1. 192.

This subject was presented to Lady Ducie as a Brazilian bird; it is, however, a native of Angola, which is well known to be situated between the rivers Dande and Coanza, on the coast of Africa. It came from Lisbon, where a great variety of curious birds are to be met with, brought thither from the Brazils, and from the Portuguese settlements on the coast of Africa.

In the formation of this bird nature, so far from being lavish, has withheld every decoration: it is neither possessed of grace or elegance, and its plumage is particularly dull and obscure; it lived but a short time in this Menagery, therefore all that could be learned of its history was, that its habits were solitary, and its character sad and sullen.

As it is my intention to give the portraits of every subject preserved in this superb collection, I have introduced this bird in the present Number by way of contrast to one so remarkable for its beauty, and the brilliancy of its plumage.

Indigo Grosbeak.

FA..K,

Summer Duck.

SUMMER DUCK.

ANAS SPONSA.	*Lin. Syst. p.* 207.
LE CANARD D'ETE'.	*Bris. Orn.* vi. *p.* 351.
LE BEAU CANARD HUPPE'.	*Buff. Ois.* ix. *p.* 243.
AMERICAN WOOD DUCK.	*Brown. Jam. p.* 481.
SUMMER DUCK.	*Catesb. Car.* i. *p.* 97. *Edw.*
	101. *Lath.* iii. 545.

THIS most elegant subject is about the size of the widgeon, and the plumage is so exceedingly beautiful, splendid, and various, that the most fortunate exertion of the artist can scarcely imitate it with success; on which account it has been named the *beautiful crested Duck*.

It is met with from New York to the West-India Islands, and also in Mexico, where it is called Yztactzon-yayauhqui, or the bird of the *various coloured head*.

It appears at New York in the latter end of February, passes the summer in Carolina, perching on the tallest trees which grow near the water (especially the deciduous cypress), and making its nest on those trees, in holes made by woodpeckers, and frequently between the forks of the branches, whence they are called the Branch, or Tree Duck. When the young ones are hatched, the old ones take them on their backs to the water, to whom the ducklings, on the least symptom of danger, closely adhere with their bill.

The natives of Louisiana ornament their calumets of peace with the neck of this bird; the flesh is much esteemed by them, and considered as a very great delicacy.

The female differs from the drake, the head is dusky brown, slightly crested, round the base of the bill, beneath the eye, chin, and throat are white, the neck and breast brown, with faint white triangular spots; the back and tail are brown; the wings brown, tinged with blue green just above the quills; across the wings is a narrow bar of white; the legs as in the male.

CHINESE DUCK.

ANAS GALERICULATA.	*Lin. Syst.* i. *p.* 206.
LA SARCELLE DE LA CHINE.	*Bris. Orn.* vi. *p.* 450.
CHINESE TEAL.	*Edw.* *p.* 102.
CHINESE DUCK.	*Lath.* iii. *p.* 548.

THIS singular and elegant bird is a native of China and Japan, where it is called *Kimnodsui*, and is held in the highest estimation by the Chinese for its beauty; it is rather less than the widgeon. The English in China call it the Mandarine Drake.

The whole plumage is a combination of the most rich and vivid colours, and to this may be added, the very remarkable singularity which distinguishes it from all other birds, of having two feathers on each side their outside webs, of an uncommon breadth; these feathers are of a bright bay, edged with black towards their points; the inner web being narrow, of a splendid blue, terminated with bay colour: these feathers appear erect when the wing is closed.

The head is adorned with a most beautiful crest of various tints, the feathers of which are very long, and fall behind the neck, the feathers of the neck are narrow, and pointed, like those of the cock, of a dull orange colour.

They are frequently exposed to sale at Canton in China, at the rate of rom six to ten dollars a couple. A pair of these scarce and valuable birds were a long time preserved in this Menagery: and though every care and attention was paid, in the hope of having them breed, it was not attended with success.

The female very much resembles that of the Summer Duck, except in having two bars of white on the wing; the breast rather more clouded with brown, and the spots rounded instead of a triangular form.

As the Chinese are not over scrupulous in their dealings, the want of success might be in consequence of having a female of the Summer Duck matched with the Chinese Drake.

Chinese Duck.

BLUE-BELLIED FINCH.

FRINGILLA BENGHALUS.	*Lin. Syst.* i. *p.* 323.
LE BENGALI.	*Bris. Orn.* iii. *p.* 303.
	Buff. Ois. iv. *p.* 92.
BLUE-BELLIED FINCH.	*Edw. pt.* 131. *female.*
	Lath. ii. *p.* 310.

ALTHOUGH the Count de Buffon has given the appropriate title of Bengal to these birds, he does not mean to infer that they are confined to that part of India, being natives of Asia and Africa. And as a number of them have been taken to the Isle of Cayenne, and there set at liberty, where they were seen to be very cheerful, lively, and disposed to perpetuate their race; he says, we may expect to see them soon naturalized in America.

A cage, containing a variety of these very lively and delicate birds, was presented to Lady Ducie, and kept in this Menagery.

The subject of this Plate had on each side a crimson crescent, placed under but rather behind the eyes; the breast, throat, belly coverts above and beneath the tail, and the tail itself, were of a delicate pale blue; all the upper part of the body and wings of a pale light grey, inclining to chesnut.

In the same cage another specimen had the lower belly and thighs the same colour as the back, and wanted the crimson spots under the eyes, which is supposed to be the characteristic of the male, as Mr. Bruce, who has seen these birds in Abyssinia, asserts that the crimson spots are wanting in the female, and that her plumage is less brilliant. The male has an agreeable warble, which he never observed in the female.

This variety came from the coast of Angola, where they are called, by the Portuguese, Azulinha, by the French Cordon bleu, and they are more frequently met with than the subject of this Plate.

PARADISE TANAGER.

Tanagra Tatao.	*Lin. Syst.* i. *p.* 315.
Le Tangara.	*Bris. Orn.* iii. *p.* 3.
Le Septicolor.	*Buff. Ois.* iv. *p.* 279.
Titmouse of Paradise.	*Edw. pl.* 349.
Paradise Tanager.	*Lath.* ii. *p.* 236.

This genus of birds, of which there are more than forty species, exclusive of varieties, are chiefly natives of the New Continent, all that have hitherto been received having come from Guiana, and other parts of South America.

They are common in the inhabited parts of Guiana, where they make their appearance generally about the middle of September, likewise at Cayenne, collecting in great flocks for the purpose of feeding on the tender half-formed fruit of a particularly large tree, where they continue about six weeks; they then take their departure elsewhere, most probably into the interior part of the country, to seek the same food, for they do not prefer any other; and wherever the trees are in bloom, these birds are certainly to be met with, and return again in April or May, at which season the fruit ripens.

The plumage of this most beautiful subject, when arrived at a mature state, is variegated with seven colours, the brilliancy of which is beyond expression, some males have the splendid red on the rump as well as the back; in others the back and rump are entirely of a gold colour, the general plumage of the female is less brilliant, and not so distinct as in the male, the lower part of the back and rump being of a dull orange.

These birds are kept in cages at the Brazils, and fed on meal and bread; they neither sing nor warble, but have only a short shrill note.

In their instinctive habits they exceedingly resemble the Sparrow, associating near the dwellings, and being particularly familiar, differing only in point of colour, and the upper mandible being slightly ridged and notched at the end.

Paradise Tanager.

GREAT BUSTARD. MALE.

OTIS TARDA.	*Lin. Syst.* i. *p.* 264.
L'OUTARDE.	*Bris. Orn.* v. *p.* 19.
	Buff. Ois. ii. *p.* 1.
BUSTARD.	*Will. Orn.* *p.* 178.
	Br. Zool. i. *No.* 98.
GREAT BUSTARD.	*Lath.* iii. *p.* 795.

THE Great Bustard is allowed to be the largest of the land fowls in our island. The male weighing from twenty to twenty-seven pounds, and measuring from three feet and a half to four feet.

The male differs from the female, not only in size and in the superior brightness of his plumage, but by the whiskers which rise on each side from the corner of the under mandible, and in his being furnished with a pouch situated in the fore part of the neck, the entrance of which is immediately under the tongue, and capable of containing seven pints of water, answer-ing the purpose of a reservoir, to supply the female while sitting, and the young birds until they are capable of providing for themselves.

Although so large a bird, they are in their wild state exceedingly timo-rous, avoiding as much as possible all intercourse with mankind; and notwithstanding their strength, are so remarkably pusillanimous as never to exert it even for their own preservation, always shrinking from attacks however contemptible the opponent, never attempting any resistance, but providing for their safety by flight. They are slow in taking wing, yet run so fast that the swiftest greyhound can only overtake them.

They inhabit most of the open countries of the southern and eastern parts of the island, from Dorsetshire to the Wolds in Yorkshire, and are frequently met with on Salisbury Plain. With us they are in the greatest numbers in autumn, sometimes in troops of fifty or more: are also common in some parts of Germany, and in Hungary, four or five hundred have been seen in a flight. They feed on grain and herbs, and likewise on large earth-worms, which appear in great quantities on the downs, where they chiefly inhabit.

A most beautiful pair of these birds were presented to Robert Child, Esq. by the present Marquis of Bath.

GREAT BUSTARD. FEMALE.

THE female Bustard differs considerably from the male in its proportions and weight, not being above half the size, and weighing from ten to twelve pounds; a greater disproportion than has been remarked in any other species.

There is likewise a difference in the plumage, the crown of the head is of a bright orange colour, marked with transverse dusky lines, the rest of the head is of a dull brown, the fore part of the neck ash colour; the hinder part and the rest of the plumage the same as that of the male, but not so brilliant.

In the season of love, the male addresses the female by strutting round her, and spreading his tail like a fan.

She does not prepare any nest, but only scrapes a hole in the ground, in the most retired part of some dry corn-field, in which she deposites two eggs, not quite so large as those of a goose, of a pale olive brown, marked with spots of a deeper shade, which she hatches after an incubation of thirty days.

When she leaves her eggs in quest of food, should any one during her absence, either touch or even breathe on them, she discovers it, and immediately abandons them. Klein remarks, that on the least appearance of danger, she will take her eggs under her wing, and transport them to a place of safety.

They are particularly attached to the place where they are bred; and as they seldom take wing but when they are closely pursued and absolutely forced to it, their greatest excursions never exceed twenty or thirty miles.

Their flesh is excellent, that of the young ones remarkably delicate.

PURPLE GALLINULE.

FULICA PORPHYRIO.	*Lin. Syst.* i. *p.* 258. 5.
LA POULE-SULTANE.	*Bris. Orn.* v. *p.* 522.
	Buff. Ois. viii. *p.* 194.
PURPLE WATER HEN.	*Edw. pl.* 87.
PURPLE GALLINULE.	*Lath. Syn.* iii. *p.* 254.

VERY few birds have more splendid or brilliant irradiation of plumage than the present subject, its beautiful feathers, when viewed in the sun, forming a combination of the richest tints of blue, purple, and green; it appears likewise to be a bird singularly disposed to domestication, being very docile, mild, easily tamed, very soon becoming familiar, and attached to its keeper. It was therefore justly considered as a valuable acquisition and an ornament to this splendid collection. It is reduced on the Plate, its length being one foot five inches.

- In Sicily these birds are bred in plenty, and very much admired for their beauty; they appear in the streets and markets, picking up the refuse of fruit and vegetables, but whether they are indigenous there, or whether they migrated originally from Africa, is uncertain.

This we know, that they abound on the coast of Barbary, in the islands of the Mediterranean; they are met with in various parts of the south of Russia, in the western parts of Siberia, and in the neighbourhood of the Caspian sea; in the cultivated rice grounds of Ghilar, in Persia, in great abundance; and in high plumage in China, the East Indies, the islands of Java and Madagascar; and they are common in the southern parts of America.

In their wild state, the female makes its nest among the reeds in March, lays three or four white eggs perfectly round, the time of incubation occupying from three to four weeks. It not only feeds on fruit, plants, and grain, but will eat fish with avidity, repeatedly dipping them in water before it swallows them: it frequently stands on one leg, and clenching its food with its toes, lifts it to its mouth with the other, in the same manner as the Parrot.

WHITE STORK.

Ardea Ciconia.
La Cicogne Blanche.

White Stork.

Lin. Syst. i. *p.* 235. 7.
Bris. Orn. v. *p.* 365.
Buff. Ois. vii. *p.* 253.
Lath. Syn. iii. *p.* 47.

In size it occupies the intermediate space between the Crane and the Heron. It inhabits most parts of Europe, except England, where it has never been seen except in four instances, and then supposed to be driven by tempestuous weather. Avoiding the extremes of heat and cold, it is not met with between the tropics. It appears in Sweden in April, retires in August, is not seen farther north than Scania, or in Russia, beyond fifty degrees north, nor to the east of Moscow.

In Lorrain, Alsace, and in Holland, these birds are in a manner half domestic, and are so far familiarized to the society of man, as to walk unconcerned about the streets, feeding on offal and filth, and clearing the fields of serpents and every noxious reptile; they likewise build on the tops of houses, forming their nest of sticks, twigs, and aquatic plants (laying never more than four eggs, oftener not more than two, of a dirty yellowish white, smaller but longer than those of a goose), on wheels and boxes provided for them by the inhabitants; where they are not only held in great veneration, but every attention is paid to their security, that no injury should be done them; and it would be almost as dangerous, in the present age, to kill a Stork in Holland, as it was in former times in Thessaly, when such a crime was expiable only by the death of the offender.

The Stork is of a mild amiable disposition (from which, as is known to the learned reader, it derives its name), particularly attentive to its young, protecting them in the moment of danger, perishing in their defence before it will forsake them.

Very much has been said by the ancients of its moral qualities; it was the emblem of temperance, conjugal fidelity, affection, and filial piety; and the law which compelled the maintenance of parents, was enacted in honour of them, and inscribed by their name.

White Stork.

Virginian Eared Owl.

VIRGINIAN EARED OWL.

Le grand Duc de Virginie.	*Bris. Orn.* i. *p.* 484.
Horned Owl.	*Ellis's Huds. Bay, p.* 40.
Great Horned Owl from Virginia.	*Edw. p.* 60.
Virginian Eared Owl.	*Lath. Syn.* i. *p.* 119.

There is little distinction between this genus and the rapacious birds, except that those commit their ravages by day, but this subject chiefly by night. The bill is short and hooked, not furnished with a cere, and both mandibles are moveable, as in the Parrot.

The nostrils are covered with bristly feathers, projecting forwards. The head and eyes are large, and during the day they are mostly shut, being unable to bear the glare of light.

The passage to the ears is large, and their sense of hearing more exquisite than that of other birds, perhaps than any known animal. Their legs and feet are for the most part clothed with feathers, down to the origin of the claws, which are much hooked, strong, and very sharp. The outermost toe is capable of being turned backwards as occasion may require, and one or more of the outermost quill feathers is serrated.

The appetite for flesh, and the disposition to plunder, are the same. This genus is subdivided into two genera, the long eared, or Horned Owl, and those with smooth heads. This is inferior in size to the Eagle Owl, not measuring more than sixteen inches.

It is common to South and North America, in Northern Asia, as far east as Kamtschatka, and almost to the North Pole; often met with at Hudson's Bay, where it frequents the woods, and builds in March in the pine tree, the nest being composed of a few sticks laid across; the eggs are two in number, of a dull white; the young fly in June.

It makes during night a most hideous noise, not unlike the outcry of a man, so that passengers beguiled by it, often lose their way in the vast forests it frequents.

GREAT RED-CRESTED COCKATOO.

Le Kakatoes 'a huppe rouge.	*Bris. Orn.* iv. *p.* 209.
	Buff. Ois. vi. *p.* 95.
Greater Cockatoo.	*Edw.* *p.* 160.
Great Red-Crested Cockatoo.	*Lath. Syn.* i. *p.* 257.

This subject is one of the largest species known, as it measures rather more than seventeen inches; the name Kakatoes, or Cockatoo, is formed from their cry.

It is distinguished from other Parrots by its size, by its white plumage, by the peculiarly incurvated shape of the bill, by the baldness of the head, and more particularly by a folding crest, near seven inches in length, the under part of a scarlet colour, inclining to orange, which it can elevate or depress at pleasure.

The bill is cinereous; the cere in which the nostrils are placed, and the orbits of the eye, are of a lead colour, the irides of a deep red. Although the general colour of the plumage is white, it is tinged on the back with shades of cream colour, on the head and breast with a soft delicate rose blush; the inner coverts of the wings with yellow, and the lateral tail feathers have their inward webs, from the base to the middle, of a sulphur-like colour; the legs and feet are lead colour, the toes black.

They inhabit the southern parts of Asia, where they seem indigenous, are likewise found in the south of India, and in all the islands of the Indian ocean, more particularly in the Moluccas under the line. In several parts of India they are in a manner domesticated, building their nests under the roofs of houses.

They seem to possess a superior understanding to that of the common Parrot, and are more docile, kind, and sincere in their attachments. This amiable disposition was particularly manifested in the subject of this Plate, for its fondness, affectionate attention, and attachment to the person who had the care of it was beyond expression.

ROSE-HEADED RING PARRAKEET,

La Perruche de Bengale.   *Bris. Orn.* iv. *p.* 348. *No.* 66.
Le petite Perruche 'a tete
coleur de rose 'a long
brins.   *Buff. Ois.* vi. *p.* 154.
Rose-headed ring-Parrakeet. *Edw. Glen. p.* 233.
                          Lath.Gen. Syn. i. *p.*339.*No.* 39.
                          Var. A.

As the genus of parrots is more numerous than any, it will not therefore
appear extraordinary that several of this family, so remarkable for the brilliancy and great variety of the plumage, as well as their agreeable and engaging manners, should form a conspicuous part of this most elegant and
superb Collection.

According to the Count de Buffon's arrangement, this subject is of the
Old Continent, and is a variety of the Blossom-headed Parrakeet. In size
it measures rather more than ten inches; the upper mandible is of a pale
yellow, with a dusky cere, in which the nostrils are placed, the under mandible black; the forehead and cheeks are of a pleasant blush, or rose colour,
which, as it approaches the hind part of the neck, gradually becomes blue.

The chin is black; a ring of the same colour encircles the neck, which
becomes narrow by degrees, and appears to divide the head from the body;
the upper part of the neck, the back, the rump, the scapulares, and coverts
of the tail, are of a pleasing green; the throat, breast, belly, thighs, and coverts under the tail, a yellowish green, some of the lesser coverts of the
wings are edged with a dull red.

The tail consists of twelve feathers, the two middle ones on the upper
surface are blue, and terminate in points, the others are tinged with green,
and gradually shorten towards the sides, the legs and claws are cinerious.
This was a very pleasing bird, extremely fond of being taken notice of, and
it never discovered the least inclination to bite or injure any stranger that
approached it.

VARIEGATED BUNTING.

EMBERIZA PRINCIPALIS.	*Lin. Syst.* i. *p.* 313. 22.
LA VEUVE D'ANGOLA.	*Bris. Orn. app. p.* 80.
LA VEUVE MOUCHETÉE.	*Buff. Ois.* iv. *p.* 165.
LONG TAIL SPARROW.	*Edw. p.* 270.
VARIEGATED BUNTING.	*Lath. Gen. Syn.* ii. *p.* 181.
	No. 18. *Var. B.*

THE subject of this plate is represented as large as life; it is a native of Angola, on the Western coast of Africa. The bill is of a lively red, short, strong, and conic, having a hard knob in the upper mandible, which enables it to break and triturate seeds and grain.

The upper part of the head and neck are of a dull orange colour, variegated with black; the back, the rump, and wings, are black, edged with dull orange; the sides of the head, the throat, the lesser coverts of the wings, the belly, and coverts under the tail, are white; the breast pale orange; the tail consists of twelve feathers, including the four long ones, which are not (as the Count de Buffon supposes) a sort of false tail of supernumerary feathers, but actually constitute a part of the true tail.

When these feathers acquire their full length, the four middle ones project considerably beyond the side ones, as expressed in the Plate; they are cast in moulting, and are quickly replaced; which, though common in most birds, is contrary to the nature of the Whidah, as that bird is often half a year without them.

It was an active, lively bird; and although it had no song, neither could it boast of any richness or brilliancy of plumage, yet the extraordinary length and form of its tail, so different from all other birds, rendered it a very pleasing and valuable acquisition to this Collection.

BLUE HEADED TURTLE.

Columba Cyanocephala.	*Lin. Syst.* i. *p.* 282. *No.* 20.
La Tourterelle de la Jamaique.	*Bris. Orn.* i. *p.* 135.
Blue-headed Turtle.	*Lath. Gen. Syn.* i. *p.* 651. *No.* 45.

The Turtle is found in every part of the Old and New Continent, even as far as the South Sea Islands. They are, like the Pigeon, subject to great variety, and though naturally more wild in their disposition, they can, with proper management, be raised and domesticated; and from the great analogy which is known to subsist between the two birds, it is not unlikely that the several varieties may result from the repeated intercourse of the Bisset, the Ring Dove, and Turtle.

The Pigeon is fond of society, attached to its companions, and faithful to its mate.—Not so the Turtle, for those that are acquainted with its manners, know it to be capricious, quarrelsome, and inconstant, notwithstanding its reputation to the contrary.

This subject is rather less than ours, yet its instincts and habits are the same, and it seems to differ only in the colour of its plumage, which may be caused by the influence of climate. A cage of these birds came from St. Lucia; they are common in Jamaica, and in the island of Cuba, where they are taken in traps to supply the market: although they had every attention and convenience, they never bred in this Menagery.

YELLOW FINCH.

FRINGILLA BUTYRACEA.	*Lyn. Syst.* i. *p.* 321. 22.
LE VERDIER DES INDES.	*Bris. Orn.* iii. *p.* 195, 55.
LE VERT BRUNET.	*Buff. Ois.* iv. *p.* 182.
INDIAN GREENFINCH.	*Edw.* ii. *p.* 84.
YELLOW-FINCH.	*Lath. Gen. Syn.* ii. *p.* 299. *No.* 68.

THIS subject is given the size of life; it is rather larger and more bulky than the Canary bird. It is found in India and at the Cape of Good Hope, at which place it is more common than the Green-finch is with us.

The bill is conic, slender towards the end, and sharp pointed; the upper mandible dusky brown, the lower lighter; the irides are hazel.

The whole upper part of the plumage is a dull olive green, a line of the same colour passes from the basis of the bill through the eye; above and beneath the eye is a line of yellow; from the angles of the mouth there is a black line in form of a mustachio.

The quills are olive green, edged with white; the whole under side, from the throat to the covert feathers under the tail, are yellow, and slightly tinged with green; the tail is a little forked, and margined with yellow-green.

It was a very bold, lively, active bird, and had a most agreeable song.

JAVA GROSBEAK.

Loxia Oryzivora.	*Lin. Syst.* i. *p.* 302. *No.* 14.
Le Gros-bec cendre de la Chine.	*Bris. Orn.* iii. *p.* 244.
Le Padda ou l'Oiseau de Riz.	*Buff. Ois.* iii. *p.* 463.
Padda, or Rice Bird.	*Edw.* i. *p.* 41.
Java Grosbeak.	*Lath.* ii. *p.* 129. *No.* 29.

Although this bird is very frequently met with at Java, and the Cape of Good Hope, there is great reason to suppose that the Europeans, in their intercourse with China and Java, had often carried these birds to that island; and that it is an inhabitant, if not a native, of China, being frequently met with in their paintings, where it is called Hung-tzoy.

It is represented on the plate of the size of life: the bill is very stout and thick for the size of the bird, of a fine red at the base, and paler towards the point, which is almost white; the head and throat black; the cheeks white; the upper part of the body, the neck, and breast, a most delicate pale ash-colour; the belly and thighs pale rose-colour; the vent, and coverts under the tail, white; the greater quill-feathers, and tail, a glossy black: legs flesh-colour.

It is very destructive to the plantations of rice, which is its principal food, and from thence called the Padda; and is remarkable for the delicacy of its plumage, which is so perfectly regular and soft, that no one feather projects beyond another, but they appear like a fine silky down, covered with a farina, or bloom.

RED AND BLUE MACCAW.

Psittacus Macao.	*Lin. Syst.* i. *p.* 139. *No.* 1.
Ara du Brasil.	*Bris. Orn.* iv. *p.* 184.
L'Ara Rouge.	*Buff. Ois.* vi. *p.* 170.
Red and Blue Maccaw.	*Edw.* iv. 158.
	Lath. Syn. i. *p.* 199. *No.* 1.

The characters which distinguish the Maccaws from other parrots, are their size, which, when in perfect feather, measures full three feet from bill to tail : the length of their tail, which is much longer than in the case of others, even in proportion to the body; and the cheeks, which are bare of feathers, being only covered with a naked membranous skin, of a whitish colour.

They inhabit Brazil, Guiana, and the warm climates of South America, and are not to be found on the old Continent. They were formerly common at St. Domingo, but in consequence of the natives having extended their plantations nearer to the mountains, they are now rarely to be met with.

They live in the woods, and prefer such as cover swampy grounds, where the palm-tree abounds, on which they feed : they sometimes assemble in flocks, but are generally found in pairs; and of all the Parrot tribe they fly the best, being known to go the distance of a league in search of ripe fruit, but always returning in the evening to their accustomed spot.

They make their nest in the holes of old decayed trees, enlarging the aperture with their bills, if too narrow; and they line the inside with feathers.

They have two broods annually, laying two eggs, spotted like those of the Partridge, and of the size of Pigeons' eggs : they sit alternately; never forsake their young, as long as their assistance is necessary; and always perch together, near the nest.

When caught young, they are easily tamed, soon become familiar, and discover great attachment to their owners; but the old birds are stubborn and mischievous.

Red and - Blue Macaw.

White Crowned Pigeon.

WHITE CROWNED PIGEON.

Columba Leucocephala.	*Lin. Syst. p.* 281. *No.* 14.
Le Pigeon de Roche de la Jamaique.	*Bris. Orn.* i. *p.* 137. *No.* 35.
Bald-pated Pigeon.	*Sloan Jam. p.* 303. *t.* 261. *f.* 2.
White-crowned Pigeon.	*Catesb. Car.* i. *pl.* 25.
	Lath. Syn. ii. *p.* 616. *No.* 5.

This subject is considered by the Count de Buffon as a variety of the wild Pigeon; it inhabits Jamaica, St. Domingo, and the Bahama Islands: they nestle and breed in holes among the rocks, and prove of infinite service to the inhabitants, who take vast numbers of them for food.

They live on the seeds of the mangrove, and wild coffee; they likewise eat the berries of *sweet-wood*; and are bitter or sweet to the taste, according to the season of the year, or more properly, according to the food they feed on; for when they meet with plenty of sweet berries they are accounted most excellent for the table.

The plumage of this bird is very delicate: the crown of the head is white, from which it takes its name; the hind neck is a combination of green, blue, and copper bronze, as viewed in different directions; the upper parts are bluish-grey, the under, lighter grey; the lesser and greater quills and tail dusky brown; the bill and legs pale red; the claws grey. Rather larger than the Turtle.

GRENADIER GROSBEAK.

LOXIA ORIX.	*Lin. Mant.* 1771. *p.* 527.
EMBERIZA ORIX.	*Lin. Syst.* i. *p.* 309. *No.* 7.
LE CARDINAL DU CAP DE	*Bris. Orn.* iii. *p.* 114. *No.* 21.
BONNE ESPERANCE.	*Buff. Ois.* iii. *p.* 496.
GRENADIER.	*Edw.* iv. *p.* 173.
GRENADIER GROSBEAK.	*Lath. Syn.* ii. *p.* 120. *No.* 16.

MONSIEUR BRISSON describes this bird under the character of the Cardinal of the Cape of Good Hope; and the Count de Buffon arranges it likewise with other foreign birds related to the Tree Sparrow; but the very accurate Latham places it, very properly, with the Grosbeaks, to which family it undoubtedly belongs.

The figure represents it the size of life, and in its summer dress, at which season, the forehead, cheeks, chin, and belly, are a downy black; the hind part of the head, the neck, the breast, lower part of the back, rump, covert feathers, above and beneath the tail, are bright orange; the back, wings, and tail, brown, each feather being dusky in the middle, and fringed with a pale rufous colour; the thighs are pale tawney; the legs flesh-colour.

In the winter it experiences an entire change; the rich black, and splendid orange, become less and less distinct, and the whole plumage is changed into dusky, brown, and tawney.

I have reason to believe, that this very rare and valuable bird does not obtain its perfect colours until the second year at soonest: as I had an opportunity of making drawings from, and examining a cage of these birds which were sent from Lisbon as a present to the late Earl of Sandwich, when at the head of the admiralty; and although there were several of them, there were scarcely two alike: in a letter which accompanied them, they were called the Portugal Bishop.

_ Blue and Yellow Macaw

BLUE AND YELLOW MACCAW.

Psittacus Ararauna.	*Lyn. Syst.* i. *p.* 159. *No.* 3.
L'Ara Bleu et Jaune du Bre-	*Bris. Orn.* iv. *p.* 193.
sil.	
L'Ara Bleu.	*Buff. Ois.* vi. *p.* 191.
Blue and Yellow Maccaw.	*Edw.* iv. *p.* 159.
	Lath. i. *p.* 204. *No.* 4.

This subject is rather less than the Red and Blue Maccaw: and nature has been lavish in clothing it with a plumage so rich and splendid, that it is scarcely within the power of the pencil to give an adequate representation of its brilliancy.

The bill is black; the nostrils are placed on a white cere; the crown of the head is green; the cheeks and throat are covered with a white skin, striped with short black feathers; the eyelids edged with black; the irides pale yellow: under the bill is a line of black, which encircles the white; the hinder part of the neck, the back, the wings, and upper side of the tail are of a rich blue; the fore part of the neck, the breast, the belly, the thighs, and coverts under the tail, are a bright yellow; the hinder part of the thighs are blue: in a word, the upper part of the bird is of a most rich glossy azure blue, the under part of a bright golden yellow colour.

The Blue Maccaw never associates with the Red, though they frequent the same situations; their voice is likewise different; neither do they articulate so plainly as the Red.

These birds are more subject to the epilepsy and cramp than any other of the Parrot tribe, more especially such as are confined; great caution should therefore be observed, that they have only a wooden perch, as it has been remarked, that such as are in the habit of resting on perches covered with iron, or tin, are invariably seized with this disorder, which often proves fatal to them.

BRAZILIAN FINCH.

Fringilla Granatina.	*Lyn. Syst.* i. *p.* 319—11.
Le Grenadin.	*Bris.* iii. *p.* 216. 67.
Red and Blue Brazilian Finch.	*Edw.* iv. *p.* 191.
Brazilian Finch.	*Lath. Syn.* ii. *p.* 316. *No.* 87.

The subject of this plate is of the size of life; it came from Brazil, where it is very much valued by the Europeans, being a lively active bird, with an agreeable song.

The bill is of a most beautiful red, perfectly conic, slender towards the end, and very sharp pointed, above the base blue; the cheeks pale violet; irides hazel, with scarlet eyelids; the head, neck, breast, and upper belly, chesnut; the throat black; the lower belly and thighs dusky; the rump and covert feathers above the tail á beautiful blue, those under the tail dusky; the wings are likewise of a dusky brown; the tail is black and cuneiform; the legs dull flesh-colour.

These birds vary very much in colour: this subject had a dusky line from the bill to the eye, in others this character is wanting: in some, the coverts, both above and beneath the tail, are violet; others have the lower belly and thighs chesnut, and the tail a reddish brown. The Count de Buffon says the Portuguese, most probably from a resemblance between the plumage of this bird and the uniform of some of their regiments, have named it the Oronooko Captain.

Cushew

CUSHEW.

Crax Pauxi.	Lin. Syst. i. p. 270. 5.
Le Hocco du Mexique.	Bris. Orn. i. p. 302. 14.
Cushew Bird.	Edw. p. 295.
Cushew Curasso.	Lath. Syn. ii. p. 696.

This subject is a native of Mexico, where it is known by the name of Pauxi. It is of the size of the Curasso, but is of a more elegant form. It differs likewise in the head, which in the Cushew is not crested, the swelling of the base of the bill is also larger, and very hard, of the size of a pear, and of a fine blue colour. The bill is red, stronger, and more hooked than the Curasso.

The upper part of the plumage is of a rich black, with intermixtures of blue and purple; the lower belly and under coverts of the tail are white, the tail is likewise tipped with white; the legs are flesh colour.

In their wild state, they perch on the highest trees, but form their nests on the ground; they are very gentle, or, more properly speaking, stupid, as they have been known to keep their station, though fired at several times.

They are chiefly found in uninhabited situations, a circumstance which accounts, in some respects, for their being so extremely rare and valuable, very few having ever been brought to Europe: the one which is described by Edwards, from the Duke of Portland's Collection, and two very fine specimens in the Osterly Menagery, being the only specimens upon record as having been ever met with in this country.

4

CHINESE QUAIL.

Tetrao Chinensis.	*Lin. Syst.* i. *p.* 277. 19.
La Caille des Philippines.	*Bris. Orn.* i. *p.* 254. *p.* 25.
La Fraise, ou Caille de la Chine.	*Buff. Ois.* ii. *p.* 478.
Chinese Quail.	*Edw. p.* 247.
	Lath. Syn. ii. *p.* 783.

This bird is represented in the Plate of its natural size; it is not only considerably less than our indigenous Quail, but it differs very materially in the colour of its plumage.

The bill is black, the fore part of the head of a bluish ash colour, all the upper part pale brown, beautifully powdered with dusky spots; the middle of each feather on the back and rump has a pale orange-colour stripe with black lines on each side; the wing feathers are brown with transverse dusky lines; the throat is black; the cheeks, and fore part of the neck white, encircled with a black band rising from each corner of the mouth, and forming a crescent on the breast: the upper part of the neck, breast, and sides, is ash colour, with transverse dusky bars; the belly, thighs, and coverts under the tail are of a reddish orange, with a line of luteous dirty white along the middle of the belly; the legs are yellow, the claws brown.

These birds are found in China and the Philippine islands, and are held in great estimation by the Chinese, who train them up to fight, as we do game cocks in England. And as they are of a very hot constitution, so much more so than any other bird, as to give rise to a proverb—*warm as a Quail*; the Chinese make a practice to carry them alive in their hands, as a security against the cold in winter.

Chinese Quail.

Shieldrake Male.

SHIELDRAKE.

Anas Tadorna.	*Lin. Syst.* i. *p.* 195. 4.
La Tadorne.	*Bris. Orn.* vi. *p.* 344. 9. *Pl.* 33.
	Buff. Ois. ix. *p.* 205.
Sheldrake, or Burrow Duck.	*Raii Syn. p.* 140. *A.* 1.
	Arct. Zool. p. 299. *D.*
Shieldrake.	*Lath. Gen. Syn.* iii. *p.* 504.

This very beautiful aquatic bird is common in many parts of England, where it continues the whole year; it breeds in rabbit burrows, in the neighbourhood of the sea, in the choice of which it is very particular, as it will enter an hundred before it fixes on one perfectly to its mind. After its taking possession, the rabbits enter no more.

The female deposites her eggs on the sand at the end of the burrow, they are in general from ten to fifteen in number; she then wraps them in soft down, which she plucks from her breast.

During the whole time of incubation, the male remains constantly on the sand bank: when the female quits her eggs to procure subsistence, the male supplies her place. The mother is particularly careful of her young, using many stratagems for their safety, when in danger, and has been known to carry them from place to place in her bill. She is rather smaller than the male, whom she resembles in colour, but the tints are less brilliant.

As they very soon become domesticated, are easily reared, and constantly retain throughout the year the charming colours of their plumage, which is so conspicuous at a distance, and has so striking an effect, not only on the water, but on land, they are most deservedly considered as a very great ornament in every collection.

This bird inhabits Northern Europe, even in the high latitude of Iceland, visits Sweden and the Orkneys in the winter, and returns in spring; is found in Asia, about the Caspian sea, and all the salt lakes of the Tartarian and Siberian deserts, and likewise in Kamtschatka: it is also met with in the Falkland Islands, and at Van Diemen's Land.

SHOVELER.

ANAS CLYPEATA.	*Lin. Syst.* i. *p.* 200. 19.
LE SOUCHET.	*Bris. Orn.* vi. *p.* 329. 6. *Pl.* 32.
	Buff. Ois. p. 191.
SHOVELER.	*Raii Syn. p.* 143. *A.* 9.
	Arct. Zool. 280. *No.* 485.
	Lath. Gen Syn. 509. 55.

This bird is less than the Wild Duck; the form of the bill being remarkably broad at the end, and spreading like a spoon, indicates its manner of living, which is on worms, aquatic insects, the larvæ of the libellula, or dragon fly, &c. The edges of the upper and lower mandibles are very much pectinated, and shut close into each other, by this means it suffers the water and mud to pass through these laminæ, and at the same time secures its food : it will likewise catch tipulæ and other flies, which pass in its way over the water, with great dexterity.

The plumage of the female is like that of the common Wild Duck, excepting that the coverts of the wings are the same as those of the Drake; she is a trifling degree smaller.

She forms her nest among tufts of rushes in the most retired and inaccessible places, laying from ten to twelve eggs, of a pale rufous colour. They run and swim as soon as they burst from the shell, at which time their bill is almost as broad as their body, and they constantly rest it on their breast, as its weight seems to oppress them. Their parents seem very much attached to them, and guard them against every appearance of danger. This bird is rather of a savage and gloomy disposition, and is not without some difficulty reconciled to domestication.

They are not very common in England, neither is it certain that they breed with us, though they are known to do so in the marshes in France; they are found in most parts of Germany, Sweden, and Norway, and breed in every latitude of Russia, especially in the north, as far as Kamtschatka; they are likewise met with in North America, at New York, and Carolina, during the winter season.

PAINTED BUNTING. MALE.

Emberiza Ciris.	*Lyn. Syst.* i. *p.* 313. 24.
Fringilla mariposa.	*Scop. ann.* i. *No.* 222.
Le Verdier de la Louisiane (dit	
le Pape.)	*Bris. Orn.* iii. *p.* 200. 55.
	Buff. Ois. iv. *p.* 176. *Pl.* 9.
Painted Finch.	*Edw.* iv. *p.* 130.
Painted Bunting.	*Arct. Zool.*
	Lath. Gen. Syn. ii. *p.* 206.

This subject inhabits the warmer parts of Canada, and is to be found as far south as Mexico, Brazil, Guiana, &c. In the summer they are met with at Carolina, but none are seen near the inhabited parts, or nearer than one hundred and fifty miles from the sea; they migrate in winter, perhaps as far as Vera Cruz, in Spanish America, where they are called by the Spaniards *Misaposa Pintada*, or the Painted Butterfly.

Nature requires time to perfect the plumage of this beautiful species; the first year it is of a plain brown; the second year the head assumes a vivid blue; the rest of the body is of a blue-green; and it is not until the third year that it fully attains the perfection of its charming tints, at which time the head and neck are of a splendid blue, the irides hazel, the orbit scarlet, the upper part of the back and scapulars variegated with different tints of green; the fore part of the neck, the breast, the belly, coverts above and beneath the tail, scarlet, inclining on the sides to orange. The lesser coverts of the wings are of a fine blue, the next light orange colour; the wings and tail dusky brown, edged with green; the legs brown.

[54]

PAINTED BUNTING. FEMALE.

INNAGRA CYANEA.   *Lin. Syst.* i. *p.* 316—6.

THIS, like the male, does not attain its full plumage until the third year, at which time the upper parts are of a dull green, and all the under parts of a yellowish green; the wings and tail are dusky brown, edged with green.

They breed in Carolina, and make their nest in the orange tree, which in particular they seem to prefer, but do not continue there during the winter.

They very soon become familiar, and there is every reason to believe, with proper care and attention, they would breed with us, as they are said to have done so in Holland. They feed on millet, canary, succory, &c. and when once properly seasoned to the climate, live eight or ten years.

These birds vary exceedingly, not only as they are some years in arriving at the height of their colour, but likewise as they moult twice a year; it will not therefore seem surprising that scarce any two should be found exactly to resemble each other.

Photo. A. Neil Burnet

BLUE-BELLIED PARROT.

Blue-bellied Parrot.   *Lath. Syn. Sup.* 59. *No.* 14. *C.*

This very beautiful Parrot is common in various parts of New Holland; and was found in great plenty in New South Wales, both at Botany Bay and at Port Jackson.

In different specimens it differs very much in the colour of its plumage, several other varieties having been met with, some of which are natives of Amboyna, and others of the Molucca islands.

It is certainly necessary, in order to prevent confusion, to preserve as much as possible the names already in use. I have therefore introduced this subject under the character of the *Blue-bellied Parrot*, on the very respectable authority of Mr. Latham, though I should have supposed that of the *Blue-headed*, under which character it was presented to Lady Ducie, would have been full as expressive and discriminating.

This very beautiful bird is fifteen inches long; the bill is orange colour; the head and throat a rich blue, inclining to purple, the middle of each feather dashed with lighter blue; the back part of the neck encircled with light yellowish-green; the breast a bright scarlet, edged with yellow; back and wings green; the scapulars are barred with scarlet, edged with yellow; the belly exhibits a most lively blue, barred with scarlet; the two middle feathers of the tail green, the others likewise green, with a bright yellow on the inner edges; the legs are lead colour.

LESSER WHITE COCKATOO.

Le Kakatoes a hupe jaune.

Lesser white Cockatoo.

Bris. Orn. iv. *p.* 206. *No.* 9.
Buff. Ois. vi. *p.* 93.
Edw. glean. t. 317.
Lath. Syn. i. 258. *No.* 64.

Of this species there are two varieties, the one much larger than the other; the subject of this plate is the least, being only fourteen inches and a half long, with a folding, pointed, sulphureous yellow crest, which the bird elevates or depresses at pleasure.

The bill and cere are blackish; the eyes are placed in a naked white skin; the irides are of a reddish colour; the whole plumage is white, with a tinge of brimstone yellow on the under parts; beneath each eye is a yellow spot; the legs and feet are black.

This very beautiful and pleasing bird was much admired by all who visited this delightful spot: its motions were uncommonly graceful, delicate, and engaging; it was fond of being taken notice of, and seemed to feel the most exquisite pleasure in being caressed; it would recline its head to the hand, and would express its joy by spreading its wings, moving its head briskly up and down, cracking its bill, and displaying its elegant crest.

It was not fond of confinement; but though it had the entire liberty of the Menagery, and would frequently climb to the top of the highest trees, it always returned at command, with much seeming satisfaction at being taken notice of.

When perched on the loftiest trees, it was much delighted at seeing a number of Crows and Daws collected around it; and whenever attacked by them (which was often the case), it never seemed the least daunted, but always drove them away, and maintained its favourite situation.

The centre tree on the sloping lawn, between the pavilion and the lake (the place of its accustomed residence), a venerable ancient oak, was in a great measure despoiled of its leaves and bark, by the indefatigable bill of this favourite creature.

Arthur Cockerton

RING PHEASANT. MALE.

THE subject of this plate was reared in this Menagery, and although these birds were formerly considered as very rare and valuable, there is every reason to believe, that in a short time they might become as abundant as the common Pheasants, from which they differ, in having a ring of a fine white colour round the neck, and also in having the colour of the plumage, particularly the feathers of the lower part of the neck and breast, more distinct, and more deeply indented, each feather appearing double at the extremity.

The bill is horn colour; the irides are yellow; the sides of the head, like that of the common Pheasant, are covered with a bright crimson carunculated bare skin, sprinkled with minute black spots, forming a point behind the eye, and stretching like the wattle of a Cock over each jaw.

But the most remarkable trait in its appearance is, two tufts of feathers, which in the breeding season rise on each side above the ears, and turning upwards, appear like two horns.

It is needless to give a particular description of the colours of the plumage, which very much resemble that of the common Pheasant, and is faithfully pourtrayed in the plate. Its size is that of a domestic fowl, measuring from two feet six inches to three feet, of which the tail is twenty inches long, consisting of eighteen feathers, the longest of which are twenty inches, the shortest not five, all having transverse bars of black on each side of the shaft, the two middle feathers having twenty-four, in some more; but the opinion that the age may be discovered from the number of cross bars on the tail, is certainly erroneous.

RING PHEASANT. FEMALE.

The female Ring Pheasant is less than the male; she likewise wants the crimson carunculated skin on the cheeks; instead of this, they are covered with feathers, and intermixed with minute warty excrescences of a pale red, hardly distinguishable: the tail is much shorter and barred; the colour of the plumage, like that of the common Pheasant, is a combination of brown, grey, rufous, and dusky colour.

In their wild state they breed but once a year, at least in our climate, and form their nest, like Partridges, on the ground, laying from twelve to fifteen eggs, smaller than those of the common Hen, laying one, in every two or three days.

When reared in our Menageries, the female selects the most retired and darkest corner of her pen, where she forms a rude unshapely nest, with leaves, straw, and whatever she can scrape together herself, which she prefers to any other materials prepared for her: the time of incubation is from twenty to twenty-five days, according to the temperature of the season, and the young as soon as hatched (like the rest of the gallinaceous tribe) run immediately on leaving the shell, following the mother like a brood of chickens.

One circumstance should be particularly adverted to, that the Hen, whilst setting, should be kept in a place remote from noise and interruption, and if somewhat under ground, she will be the less affected by thunder storms, or any sudden variation of the weather.

BALD EAGLE.

Falco leucocephalus.	*Lyn. Syst.* i. *p.* 124. *No.* 3.
L'aigle a tete blanche.	*Bris. Orn.* i. *p.* 422. *No.* 2.
Le Pygargue.	*Buff. Ois.* i. *p.* 99.
White headed Eagle.	*Arct. Zool.* i. *p.* 228. *No.* 89.
Bald Eagle.	*Lath. Gen. Syn.* i. *p.* 29. *No.* 3.

This is a bird of singular courage and spirit; it is three feet three inches long, and weighs nine pounds. It inhabits the northern parts of Europe and Asia, it is likewise met with at Nootka Sound, and Kamtschatka, and is common in North America. It is called *Wapaw-Ertequam-Meekesue* at Hudson's Bay.

It preys on lambs, fawns, pigs, and fish; the latter it cannot procure for itself, but fixing on a convenient situation for watching the motion of the Osprey, the moment it perceives that bird has, by diving, seized a fish, it does not cease pursuing its prey, till the Osprey, of which it is the terror, drops its prize, which the Eagle seizes with astonishing dexterity in the air, before it can fall to the ground; and constantly attends the sportsman, and snatches up the game he has shot, before he can reach it, or reload his gun.

It builds in the highest trees, preferring decayed cypresses, or pines, which hang over the sea, or some great river; it associates with Ospreys, Herons, and other birds, and forms its nest of a large size, upwards of six feet in width. Their nests are composed of sticks and grass, and are so numerous as to represent a rookery. It breeds very often, laying again under their callow brood, whose warmth assists to hatch the eggs.

It is clear that this bird either is, or was formerly, an inhabitant of Asia Minor. It is with a reference to its uncommon appearance, as well as to the customs of the Jews in their mourning, that the figurative expressions of an ancient Prophet may be understood: " Make thee *bald*, and poll thee for thy delicate children : *enlarge thy baldness as the Eagle*, for they are gone into captivity from thee!" Micah, ch. i. ver. 16.

BLACK BELLIED GROSBEAK.

BLACK BELLIED GROSBEAK.   *Brown's Ill. p. 58. pl. 24.*

A CAGE of these birds was presented to the late Earl of Sandwich, when at the head of the Admiralty, under the character of the Ghillan Sparrow, the note which accompanied them, described them as coming from the Brazils. His Lordship presented a pair to Robert Child, Esq. from which this drawing was taken.

It is represented the size of life. The bill is dusky, the head, the hind part of the neck, the back, wings, and tail, are brown, each feather being dusky in the middle, and edged with light brown. The chin and belly a very rich black, with a purple gloss; the breast, sides, and coverts under the tail, are bright yellow; this is its summer dress. In the winter it becomes entirely brown. The legs are tawny.

Black Bellied Grosbeak?

Spoonbill.

SPOONBILL.

PLATALEA LEUCORODIA.	*Lin. Syst.* i. *p.* 231. 1.
LA SPATULE.	*Briss. Orn.* v. *p.* 352.
	Buff. Ois. vii. *p.* 440. *pl.* 24.
WHITE SPOONBILL.	*Lath.* iii. *p.* 13. 1.

THE body of this very extraordinary subject is nearly as large as that of the Heron, but its neck is not so long, nor its legs so tall. The bill is long, broad, flat, and thin; the end widens into a roundish form not unlike a spoon, from whence it takes its name. The whole plumage is generally white; in some specimens, indeed, the quills are tipped with black, but this does not denote, as is supposed, a difference of sex, as it occurs both in male and female.

They are met with from the Ferro Isles, near Iceland, to the Cape of Good Hope. At Sevenhuys, near Leyden, they were once in great plenty, annually breeding in a wood near that place; they are seldom seen in England, except when driven there by accident. A whole flock of them was seen in the marshes near Yarmouth, in the month of April, 1774.

They build their nest on high trees near the sea, laying four white eggs, powdered with pale red, and are (like the Rook) very clamorous during the breeding season. Their food is fish, which it is said they have the art of taking from other aquatic birds, frightening them by clattering the bill until they drop their prey; they likewise devour frogs and snakes, and in case of necessity feed on aquatic plants, and the roots of reeds.

They are migratory birds, retiring to the warmer parts on the approach of winter. Their flesh is eaten by some, and is said to have the flavour of a goose; and the young birds, taken before they are able to fly, have in particular been esteemed good food.

SCARLET IBIS.

TANTALUS RUBER.

LE COURTY ROUGE DU BRASIL.

SCARLET IBIS.

Lin. Syst. i. *p.* 241. *No.* 5.
Bris. Orn. v. *p.* 344. 12.
Buff. Ois. viii. *p.* 35.
Arct. Zool. ii. *p.* 161.
Lath. iii. *p.* 1. 106.

This very beautiful bird is somewhat larger than the English Curlew, and is met with in most parts of America, within the torrid zone, but very seldom beyond the northern tropic.

It has a slender incurvated bill, in length between six and seven inches, of a pale red colour; and the whole plumage is of the richest and most glowing scarlet imaginable, except four of the outer prime quills, which are of a glossy black at the ends; the shafts of the quills and tail are white; the legs pale red.

These birds generally frequent the low marshy grounds contiguous to the sea, and the sides of great rivers; and in the state of nature live on the small fry of fish, or shell-fish, and on the insects which they find when the sea retires from the shore.

They are easily domesticated, so as to be completely attached to the poultry yard, in which state they soon become familiar, and will feed from the hand.

They shew great courage in attacking the fowls, soon become masters of the menagery, and will even oppose themselves to cats. They frequently perch on the trees in great numbers, but lay their eggs on the ground; the young, when first hatched, are of a dusky colour; they then grow grey, and afterwards white; but it is not until the third year that the brilliant scarlet becomes complete. The flesh is excellent.

Scarlet Ibis

WEAVER ORIOLE.

Le Cap-More.   *Buff. Ois.* iii. *p.* 226.
Troupcale male du Senegal.   *Pl. ent.* 375.

This bird is rather larger than the Baltimore Oriole, described in the 16th Plate of this Work. It is a native of Africa; and the subject of this Plate, which I have reason to think was the first that was ever brought to England, came with several other birds from Senegal.

It is not until they reach their second year that these birds attain their perfect plumage, at which period, in the spring, the head, which was before yellow, becomes of a rich brown, which when viewed in the sun appears to have a brown golden gloss, in the shape of a friar's cowl, from whence it takes the name of Cap More, or Capuchin Mordore. In moulting this cowl disappears, leaving the head of its yellow colour, which returns again regularly every spring.

The Count de Buffon describes two of these birds which interwove the stalks of groundsel, chickweed, pimpernell, &c. with which they were fed in the wires of their cage. As this was understood as an indication of a desire to breed, they were supplied with proper materials, with which they soon formed a nest sufficiently capacious to conceal one of the birds entirely; but the labour of one day was destroyed in the next; which proves that in a state of nature the fabrication of the nest is not the production of an individual, but the joint efforts of both male and female, and is most probably finished by the last.

On some sewing silk being put into the cage, it interwove the wire so as to render that side on which it worked from being seen through, and it was remarked that it gave the preference to green and yellow to any other colour.

The note of this bird is sprightly, but too shrill to be agreeable.

DOMINICAN GROSBEAK.

Loxia Dominicana.　　　　　*Lyn. Syst.* i. *p.* 301. *No.* 8.
Le Gros-bec du Brasil.　　　*Bris. Orn.* iii. *p.* 246. *No.* 11.
Dominican Cardinal.　　　　*Edw.*　*pl.* 127.
Dominican Grosbeak.　　　　*Lath. Gen. Syn.* ii. *p.* 1—21.

Tнıs beautiful bird is a native of South America; the Count du Buffon describes it from Margrave, under the Brazilian name *Paroare*. Edwards as the Dominican Cardinal, from the circumstance of its head being red, and its body black and white.

Its length is seven inches; the head, throat, and fore part of the neck, is of a full rich red; the hind part of the neck of a deep ash-colour, inter-mixed with white; the coverts of the wings, the back, the rump, and co-verts of the tail and scapulars, are gray, mixed with a few black spots; the sides of the neck, the breast, belly, thighs, and vent, are white; the quills are dusky, edged with white; the tail is black, and is near three inches long; the bill is strong, the upper mandible brown, the under of a pale flesh-colour; the irides blue; the legs cinereous.

Several of these birds have been in this Collection; and as there was not the least variation in their plumage, it was impossible to distinguish the difference of sex; though the Count du Buffon describes the female as hav-ing the head of a yellow orange, sprinkled with reddish points.

They had no song, nor ever made any attempt at one.

THE HOOPOE.

Upupa epops.	*Lin. Syst.* i. *p.* 185. *No.* 1.
Le Hupe ou Puput.	*Briss. Orn.* ii. *p.* 455. *No.* 1.
	Buff. Ois. vi. *p.* 439.
Hoop, or Hoopoe.	*Will. Orn. p.* 145.
Common Hoopoe.	*Lath. Gen. Syst.* i. *p.* 687.

Tiis very singular bird is well known by its crest, which consists of a double row of parallel feathers ; the highest of these is about two inches long, and in its natural position is reclined backwards; but whenever the bird is pleased, or in a state of agitation, becomes erect, and spreads itself in the form of a fan, as represented in the Plate.

The Hoopoe is twelve inches in length, and weighs about three ounces. It inhabits Europe as far as Sweden, where it is called Narfogel, or Soldier Bird, from its note ; as it runs on the ground it frequently utters the note opp, opp, opp, thrice reiterated; and then hastening most swiftly to another spot, it repeats the same cry, resembling the word opp, which in the Swedish language signifies to " arms!" hence the common people believe the appearance of this bird to be an omen of war.

It is likewise scattered through Asia and Africa, being met with at the Cape of Good Hope, Ceylon, and Java. At the same time it has frequently been found in England, and has been known to breed in this country.

It feeds, in a state of nature, on insects ; and it is from the exuviæ of the large beetles and other insects with which the nest is crowded that it has in general so horrible a smell. It breeds in preference amidst putrid carcasses, in hollow trees, holes in the walls, and sometimes on the ground, laying from two to seven eggs, of an ash-colour.

This very beautiful bird had been seen several times in this Menagery ; every scheme that could be thought of was put in practice to take it alive, but without effect; and as it was observed that its visits were less frequent, and that it became more cautious, it was at last shot by the Menagery keeper; and by the late Mr. Child's particular orders this drawing was made for his Collection.

RED HEADED GUINEA PARRAKEET.

PSITTACUS PUTTARCUS.	*Lin. Syst.* i. *p.* 149.
LE PETITE PERRUCHE DE GUINEA.	*Bris. Orn.* iv. *p.* 387.
LE PERRUCHE À TETE ROUGE OU LE MOINEAU DE GUINEA.	*Buff. Ois.* vi. *p.* 165.
LITTLE RED-HEADED PARRAKEET, OR GUINEA SPARROW.	*Edw.* 237.
RED-HEADED PARRAKEET.	*Lath. Gen. Syn.* i. *p.* 309.

THIS small species is about the size of a Lark, and is five inches and a half long. It is dispersed over almost all the southern climates, for it is found in Ethiopia, India, Java, and is also met with at Surinam. These birds are very common in Guinea, and are frequently exported from Africa in great numbers to Europe, although not above one in ten survives the passage. Those, however, who escape this first danger, often live many years in these latitudes, after their arrival.

They are more admired for the brilliancy of their plumage, their tameness, and docility, than for any thing else; for though they can, in a certain degree, imitate the manners of other birds, they do not talk, and their note, or rather their scream, is very harsh and discordant.

They are remarkable for their affection for each other, and seem wretched in case of a minute's separation. If one of a pair is sick, the other immediately becomes melancholy; and should the one die, its associate is hardly ever known to survive.

The female differs from the male in the circumstance of having less vivid colours; the red is much paler, and inclines to orange, and the ridge of the wing is yellow, which in the male is blue.

BLACK GROUS. MALE.

TETRAO TETRIX.	*Lin. Syst.* i. *p.* 272. 2.
LE COQ DE BRUYERES À QUEUE FOURCHE'.	*Briss. Orn.* i. *p.* 186. 2.
LE PETIT TETRAS, OU COQ DE BRUYERE.	*Buff. Ois.* ii. *p.* 210.
BLACK COCK, BLACK GAME.	*Raii Syn. p.* 53. *A.* 2. *Will. Orn. p.* 173.
BLACK GROUS.	*Lath.* ii. *p.* 733.

THIS bird is rather larger than a fowl, being almost two feet in length: a full grown cock will weigh nearly four pounds. The seven exterior feathers of the tail curve outwards, and the ends are square, and appear as if cut off; the middle ones are much shorter, making the tail appear forked: is fond of woodland and mountainous situations; perches like the pheasant. It feeds on acorns, bramble-berries, and bilberries. In the summer it quits the hills and descends into the plains, and feeds on grain; in the winter, on the tops of the heath.

They never pair, but in the spring the male gets upon some eminence, crows, and claps his wings; on which signal all the females within hearing assemble: he then makes choice of two or three hens, to which he particularly attaches himself.

They are common in all the northern parts of Great Britain, but more especially in Scotland and Wales, on the fells of Cumberland and Westmorland, the moors in Yorkshire, Staffordshire, the New Forest of Hampshire, and likewise in Sussex.

Every attention was paid to this pair, which were preserved in this menagery in order to their breeding, but without effect.

BLACK GROUS. FEMALE.

THE colour of this bird differs so materially from that of the male, that Gesner was induced to consider it as a distinct species. Her tail is likewise less forked, and she is considerably less, being only one foot six inches in length, and weighing about two pounds.

As soon as she becomes impregnated, she seeks the most sequestered spot, where she retires. She is at very little pains in forming her nest; laying six or eight eggs, according to some, and even from twelve to twenty, according to the Count de Buffon, of a dull yellowish white colour, marked with small ferruginous specks, and towards the smaller end with larger blotches of the same colour.

As soon as the chickens are twelve or fourteen days old, they flap their wings and attempt to fly; but it is five or six weeks before they are able to rise from the ground, and then they perch on the trees with their mother: the young males quit their mother in the beginning of winter, and keep in flocks of seven or eight till spring; during that time they inhabit the woods. They are very quarrelsome, fighting most furiously like game-cocks with each other, until the vanquished are put to flight; and at that time they are so inattentive to their own safety, that two or three have frequently been killed at one shot.

GUERNSEY PARTRIDGE.

TETRAO RUFUS.	*Lin. Syst.* i. *p.* 276. 12.
LA PERDRIX ROUGE.	*Bris. Orn.* i. *p.* 236. 10.
	Buff. Ois. i. *p.* 431. *pl.* 15.
PERDRIX RUFTA.	*Raii Syn. p.* 57. *A.* 5.
GUERNSEY PARTRIDGE.	*Lath.* ii. *p.* 768.

THESE birds are rather larger than the common partridge, and are found in various parts of Europe, Asia, and Africa; they are common in France, Italy, and the islands of Madeira, Guernsey, and Jersey: where their flesh is very much esteemed. In France they are made into pies, and considered as a very great delicacy, being sent as presents to every part of Europe.

From their being some time met with at large in this country, it is reasonable to think they might, with proper care and attention, become naturalized, and breed here: but although several attempts have been made for that purpose, by turning out several brace at the proper season, the experiment hitherto has always failed; and as they have been known to succeed in this menagery, it is supposed our climate is either too moist, or too chilly for this bird in a wild state.

The female differs from the male, in wanting the blunt knob, or spur, behind the leg. It is a very beautiful bird, and most deserving a place in every curious collection.

In the Isle of Cyprus, the red partridge is often used, as we do game-cocks, for the irrational amusement of butchering each other.

BLUE HEADED TANAGER.

Le Tangara varie `a tete bleue de Cayenne.	*Bris. Orn. Suppl. p.* 62.
Le Tricolor.	*Buff. Ois.* iv. *p.* 276.
Blue headed Tanager.	*Lath.* ii. *p.* 235. *Var. A.*

This very beautiful bird is of the same family of the Paradise Tanager, described in page 32 of this work. Whether it is the male or female of the green-headed Tanager, is not ascertained; but there is not the least doubt of their being of the same species, and differing only in sex.

This subject is represented of the size of life. It came from Cayenne, an island of South America, situated at the mouth of the river Amazons, called by the French, Equinoctial France, from its being nearly under the Equinoctial line. That described by Mons. Brisson, from the cabinet of Madame de Bourjourdain, came likewise from Cayenne. But the bird which the Count de Buffon describes, from the cabinet of Mons. Aubri, rector of St. Louis, is said to have come from the Straits of Magellan; but it is not probable this bird should inhabit the torrid climate of Cayenne, and likewise the frozen tracts of Patagonia.

NONPAREIL PARROT.

Psittacus Eximius.   *Lin. Syst. Nat. p.* 139.

The subject of this Plate is a native of New Holland, a country from whence has been received a greater number of birds, of superior brilliancy and varied plumage, than from any other part of the known world.

It is nearly fifteen inches in length, and in form very much resembles the Pennantian Parrot. It is a very lively, active, docile bird, fond of being taken notice of, which it returns by grateful and tender caresses: all its actions, all its motions, so different from most of the parrot tribe, have a gentleness and grace, which add new charms to its beautiful plumage: and I consider it as one of the most beautiful birds that ever came under my inspection.

COWRY GROSBEAK.

LOXIA PUNCTULARIA.	*Lin. Syst. p.* 302. *No.* 18.
LE GROS-BEC TACHETE' DE JAVA.	*Briss. Orn.* iii. *p.* 238.
GOWRY BIRD.	*Edw. pl.* 40.
COWRY GROSBEAK.	*Lath. Gen. Syn.* ii. *p.* 142.
	No. 50.

THIS bird is a native of the island of Java, and has been described by Edwards under the name of the Cowrie Bird, because its ordinary price is a cowrie, one of the small shells which pass in India for money.

By Brisson it is called the Spotted Grosbeak of Java, and by Albin, the Chinese Sparrow. It is exhibited on the Plate as large as life, and is a very pleasing lively bird.

The Count de Buffon considers those of this species whose bellies are spotted, to be the males, and those which are not spotted, females.

VARIEGATED PHEASANT.

Le Faisan panaché'.	*Briss. Orn.* i. *p.* 267.
Le Faisan varié'.	*Buff. Ois.* ii. *p.* 352.
Variegated Pheasant.	*Lath. Gen. Syn.* ii. *p.* 716. *No.* 4.

This subject is a mixed breed, between the white and common Pheasant, which it exactly resembles in size and shape, the only difference being in its colour, which is of a pure white, variegated with irregular blotches of chesnut, &c. as expressed in the Plate.

Its habits are likewise the same, though they are considered as being more delicate, and of course less able to bear the severity of the winter: they are nevertheless much valued, as forming a pleasing variety in every collection, though they are not so proper for propagation.

ANGOLA FINCH.

La Linotte d'Angola.	*Briss. Orn. Suppl.* 81.
La Vengoline.	*Buff. Ois.* iv. *p.* 80.
Linnet from Angola.	*Edw. pl.* 129.
Angola Finch.	*Lath. Gen. Syn.* ii. *p.* 309.
	No 78.

This bird is a native of Angola, a kingdom in Africa, situated between the equinoctial line, and eighteen degrees of south latitude. In size and character it very much resembles our linnet.

The male is called Nigral, or Tobaque, by the Portuguese; the female, Benguelinha; and it is ranked among the finest warblers of that country. It is also mentioned by the Honourable Daines Barrington, in the Philosophical Transactions, under the character of the Vengoline, as excelling in point of song, *all the birds of Asia, Africa, and America,* the mocking bird only excepted.

Edwards has described it, and is inclined to think his subject is the female of the Nigral; but as the females in general seldom have any song, and this bird is allowed to have a remarkable fine one, it is very natural to suppose that the bird described by him was a male, although it had not arrived at its full plumage.

Icterus Oriole

ICTERIC ORIOLE.

ORIOLUS ICTERUS.　　　　　*Lin. Syst.* i. *p.* 161. *No.* 4.
LE TROUPEALE.　　　　　　*Briss. Orn.* ii. *p.* 86. *pl.* 8. *fig.* 1.
BANANA BIRD, FROM JAMAICA. *Albin.* ii. *pl.* 40.
ICTERIC ORIOLE.　　　　　　*Lath. Gen. Syn.* ii. *p.* 424.

THE American Continent is the native region of this species, which is near ten inches long from the point of the bill to the end of the tail. It is met with in Carolina, the Brazils, and in all the Caribbee Islands.

It is nearly related to the Stare, and may very properly be considered as the Stare of the New World. The instincts and habits are the same; except in the method of building their nests, which is the most curious part of the history of the Troupeale.

These nests are in the form of a cylinder, and are suspended from the extremity of the branch of a tree; so that the young are not only continually rocked by the wind, but are likewise secure from the attempts of land animals, and especially snakes, which would otherwise destroy the young birds.

These birds in their wild state are very agile, and irritable, and are so bold as even to attack men: when reclaimed, they are of a very docile, social disposition; so much so, that in America they are kept in houses for the express purpose of killing flies and other insects.

The bill of these birds is long and pointed, and seems to have no constant colour; in some it is grey, in others of a horn colour, and black : the legs are likewise subject to the same variation.

WHITE HEADED GROSBEAK.

Loxia Maca.	*Lyn. Syst.* i. *p.* 301. *No.* 11.
Le Maca de la Chine.	*Briss. Orn.* iii. *p.* 212. *No.* 65. *pl.* 9. *fig.* 2.
Le Maian.	*Buff. Ois.* iv. *p.* 107. *pl.* 3.
Malacca Grosbeak.	*Edw. pl.* 306. *fig.* 1.
White headed Grosbeak.	*Lath. Gen. Syn.* ii. *p.* 2. 151.

The Count de Buffon has described this bird under the character of Maca, and considers it as a variety of the Grosbeak of Senegal. His subject differs from that which was preserved in the Osterley Menagery, in having the *head* as well as the *belly* black.

It inhabits Malacca and China, and most probably all the intermediate countries. It differs materially from the American Maca, or Cuba Finch, which it approaches most nearly; but there are distinctions which authorize our classing it as a separate species.

The head and neck are of a dusky white; all the upper part of the body, wings, and tail, are of a reddish chesnut colour; the breast pale chesnut; the belly and vent dusky; the bill and legs lead colour.

Although nature has not adorned this bird with any of its choicest, or richest colours, and though in point of brilliancy of plumage it fell far short of many of its feathered companions, yet it formed a very pleasing variety in this most elegant Collection, and was a very lively and familiar bird.

BLUE JAY.

Corvus cristatus. Lin. Syst. i. p. 157. No. 8.
Le Geay bleu de Canada. Briss. Orn. ii. p. 55. No. 4.
Blue Jay. Edw. pl. 239.
Lath. Gen. Syn. i. p. 386.

This very beautiful bird is a native of North America, it is met with at Albany, and as far south as Carolina.

In size it is less than the common Jay, which it very much resembles in its actions, and petulance of character; though its cry is less harsh and discordant. The feathers on the crown of the head are long, which it can raise at pleasure into a crest larger than the European Jay: its plumage is a pleasing intermixture of blue, white, black, and purple, which renders it one of the most beautiful birds in this Collection.

It feeds on fruits and berries, and commonly spoils more than it eats. It is particularly fond of the bay-leaved smilax, and of maize. These birds, frequently uniting in flocks of twenty thousand at least, soon lay waste a field of ten or twelve acres; on which account they are considered the most destructive of the feathered race in that country.

They build in swamps, along with the red winged Oriole, in the month of May: laying five or six eggs, of a dusky olive colour, with ferruginous spots.

BEARDED TITMOUSE. MALE.

Parus Biarmicus.	*Lin. Syst.* i. *p.* 340. *No.* 12.
La mesange barbue, ou le	
Moustache.	*Briss. Orn.* iii. *p.* 567. *No.* 12.
Least Butcher-bird.	*Edw. pl.* 55.
Bearded Titmouse.	*Lath. Gen. Syn.* ii. *p.* 552.

The Countess of Albemarle brought a cage of these birds from Denmark, and some having made their escape from confinement, the Count de Buffon supposes, that from this circumstance a colony was found in England.

They are frequently met with in the marshes between Erith and London, in the like situations near Gloucester, and among the reeds near Cowbet in Lancashire; in all which places there is every reason to suppose they propagate their species, as it is known for certain that they remain with us the whole year; for which reason, Mr. Latham is inclined to think that they have been indigenous *ab origine*, and that it is merely owing to their frequenting the reed beds (which being overflowed by every tide, are inaccessible to us), which accounts for the habits of these birds being so little known to English naturalists.

A characteristic feature of the male, is a tuft of pretty long black feathers on each side of the head, which rise a little above the eyes, and turning downwards fall on the neck, where ending in a point, they exhibit the resemblance of mustaches or whiskers; and to this singularity the name of the bird in different countries may be traced.

The female is rather smaller, and differs from the male in not having the black feathers on the sides of the neck; the crown of the head is of a ferruginous colour, spotted with black; the outmost feathers of the tail are black, the ends white.

Bearded Titmouse
Male

The Bronze winged Pigeon.

GREEN WINGED PIGEON.

COLUMBA INDICA.	*Lin. Syst.* i. *p.* 284. *No.* 29.
LE PIGEON RAMIER D'AM-	*Briss. Orn.* i. *p.* 150. *No.* 42. *pl.*
BOINE.	15. *fig.* 1.
GREEN-WINGED DOVE.	*Edw.* i. *pl.* 14.

THE Pigeon, from having a very powerful wing and a well supported flight, can easily perform very distant journies. It is, therefore, more generally dispersed than any other species; for most of our wild and tame sorts are to be met with in every climate; and they are so little affected either by heat or cold, that the wild Pigeon is diffused through the whole extent of both Continents.

A pair of these very beautiful birds were for some time preserved in this Menagery. They are natives of Amboyna, an island in the East Indies, the chief of the Moluccas; and are about the size of the Turtle, though different in the distribution of its colours, which may be attributed to the effect of a hot climate.

The Count de Buffon seems to consider it as a variety of the European Pigeon, which is found in Mexico, Martinico, Cayenne, Carolina, and Jamaica; that is, in all the warmer, and temperate climates of the West Indies; and in the East, from Amboyna to the Philippines.

FASCIATED GROSBEAK.

Loxia Fasciata.

THIS bird is a native of Africa, and was presented to Lady Ducie under the character of the *Cut-throat Sparrow*. It has been described by Browne, from a bird in the late Mr. Tunstall's Collection, and was at that time considered as a very rare species; but they are not so at present, as I have seen several in the late Earl of Sandwich's magnificent Collection, as well as at Osterley.

It is represented the size of life; the bill is bluish grey; the head, the hinder part of the neck, the back, and coverts of the wings, pale brown, marked with some circular black lines: the throat and cheeks white, bounded beneath with a rich band of crimson, edged with a black line; the breast and belly of a pale brown, faintly marked with semicircular lines of darker brown. The legs are flesh colour.

3 Fasciated Grosbeak.

CARDINAL CROSBEAK.

Loxia Cardinales.	*Lin. Syst. p.* 300. *No.* 5.
Le Grosbec de Virginie.	*Briss. Orn.* iii. *p.* 253. *No.* 17
Virginia Nightingale.	*Ray Syn. p.* 85. *A.* 3.
La Cardinal Hupé,	*De Buff.* iii. *p.* 458.
Cardinal Grosbeak.	*Lath. Gen. Syn.* ii. *p.* 118. *Arct. Zool.*

This species measures upwards of eight inches in length, and is a native of the temperate climates of America, inhabiting the country from Newfoundland to Louisiana. It arrives about the beginning of April in New York and the Jerseys, and frequents the Magnolia swamps during the summer; departing again for Carolina at the commencement of autumn.

They are very hardy, familiar, and docile birds, on which account attempts have been made to breed them in cages, but without success; as in a state of confinement, the male and female are at such enmity that they frequently kill one or other. It is true, that a relation of the late Mrs. Tunstall had a pair which built in an orange tree, in the Aviary; but while the hen was sitting, a high wind blew down the nest, whereby the eggs were broken: young birds were found in them. This, however, is a solitary instance in the history of the Cardinal Grosbeak, as connected with this country.

In spring, and during most part of the summer they set warbling, in the morning, on the highest trees; and their song being remarkably fine, they have very deservedly obtained the name of Nightingale.

They are fond of *maize* and *buckwheat*, and will get together hoards of that grain for a winter provision, which they very artfully conceal by means of leaves and small twigs, leaving only an aperture for entrance into their magazines. The female is likewise crested, but her colours are not so splendid as those of the male, being of a reddish brown.

SAPPHIRE CROWNED PARRAKEET.

Psittacus galgulus.	*Lin. Syst.* i. *p.* 150. *No.* 46.
La petite Perruche de Ma-	*Briss. Orn.* iv. *p.* 386. *No.* 14.
lacca.	
La Perruche a tete bleue.	*Buff. Ois.* vi. *p.* 163.
Sapphire crowned Parrakeet.	*Edw.* vii. *p.* 293. *fig.* 2.
	Lath. Gen. Syn. i. *p.* 312. *Var. A.*
Petite Perruche de L'Isle	*Son. Voy. p.* 76. *t.* 33. *lowest fig.*
de Lucon.	

This subject is the size of life. Edwards has described this bird as coming from Sumatra; and according to Sonnerat, it is found in the island of Luconia, or Manilla, the chief of the Philippine Islands.

The crown of the head of this bird is of a rich sapphire blue; and round the hinder part of the neck is an orange half collar; the breast and rump are red, the rest of the plumage green.

Osbeck says, he met with this bird in the island of Java, where the natives call it *Parkicki:* if confined in a cage it very seldom whistles, and generally becomes quite sullen; it sleeps suspended by one foot, and frequently hangs itself by its feet in such manner that the beak is turned towards the bottom of the cage, seldom changing this strange situation.

In their wild state they are particularly fond of the fresh juice of the cocoa tree, called *Callou:* this subject fed on boiled rice.

Sapphire Crowned Parrakeet.

YELLOW HEADED ORIOLE.

Oriolus icterocephalus.	*Lin. Syst.* i. *p.* 163. *No.* 16.
Le Carouge a` tete jaune de	
Cayenne.	*Bris. om.* ii. *p.* 124. *No.* 27.
Les Coiffes jaunes.	*Buff. ois.* iii. *p.* 217. 250.
Yellow headed Starling.	*Edw.* *p.* 323.
	Lath. Gen. Syn. i. *p.* 441. 30.

This bird is a native of Cayenne, in South America, and is described by Edwards under the name of the Yellow Starling, and also by the Count de Buffon, as the Cayenne Bonana, with a black plumage, and a sort of cowl, or coif, on the head, throat, and forepart of the neck, descending lower before than behind, of a most splendid beautiful yellow colour.

The bill is blackish, strait, conical, very sharp pointed; the edges cultrated, inclining inwards, and the mandibles of equal length; the base of which is covered with short black feathers, extending from the nostrils to the eyes, as expressed in the Plate. The legs and claws are tawny, and its length is seven inches; its extent, eleven.

. These birds are gregarious, frugivorous, and granivorous; and are considered as very crafty and voracious; very numerous, and forming pensile nests.

JACARINA TANAGER.

TANAGRA JACARINA.	*Lin. Syst.* i. *p.* 314. *No.* 4.
LE TANGARA NOIR DU BRESIL.	*Bris. om.* iii. *p.* 28. *No.* 16.
LE JACARINI.	*Buff. ois.* iv. *p.* 293.
JACARINI.	*Edw.* *p.* 306.
JACARINI TANAGER.	*Lath. Gen. Syn.* iii. *p.* 238. 34.

THIS subject is the male, and is represented on the plate of the size of life. It is a native of the Brazils, and is called Jacarini; by the Portugueze *Negretto*, from the general colour of the plumage being black, though exceedingly glossy, like burnished steel, with green and blue reflections, as viewed in different lights.

The female differs so much in the colour of her plumage, being entirely grey, that she may be taken for a different species.

This bird is very common in Guiana, in open situations, more especially such as are newly cultivated, though never met with in large forests; it lodges in low trees, especially the coffee-tree.

The male is distinguished by a very singular circumstance; that of springing upwards from the branch it first perched on, and then falls on the same spot, first resting on one foot, and then on the other, alternately, at each leap, when it takes its flight to another bush, repeating the same exercise.

This is the manner in which he addresses the female, as each bound is attended with a plaintive cry expressive of pleasure, and likewise by an expansion of the tail.

BLUE GROSBEAK.

Loxia Cyanca.	*Lin. Syst.* i. *p.* 303.
Le Gros-bec blue d'Angola.	*Bris. om. app. p.* 88. 19.
Blue Grosbeak from Angola.	*Edw.* *p.* 125.
Blue Grosbeak.	*Lath. Gen. Syn.* ii. *p.* 117. 11.
	Vas. B.

This Plate represents the bird as large as life. Although it came from Lisbon, it is supposed to be a native of Angola, in Africa, a country situated between the river *Dande* and *Coanza*, in Congo, and called *Azulam* by the Portugueze, who have several colonies and settlements there.

The bill of this bird is very strong, double convex, very thick at its base, and of a lead colour; it is surrounded with black feathers, which reach on each side as far as the eye, and descend under the chin.

The irides are of a dark hazel; and the general plumage is of a fine deep blue, except the wings and tail, which are black; the legs are very strong and black.

GREEN GOLD FINCH.

Fringella melba. Lin. Syst. i. p. 319. 8.
Le Chardonneret verd. Bris. om. App. p. 70.
Le Chardonneret verd, ou le
    Maracaxao. Buff. ois. iv. p. 211.
Green Gold Finch. Edw. p. 272.
    Lath. Gen. Syn. ii. p. 286. 52.

This very pleasing bird inhabits China, and also is met with at the Brazils, where it is called *Maracaxao*, by the Portugueze.

In size it is rather larger than the common Gold Finch, which it very much resembles in form and manners.

The bill is flesh colour, conical, slender towards the end, and very sharp pointed; the forepart of the head and throat, and a little behind the eyes, is a rich red, or scarlet; the space between the bill and eye is of a bluish ash colour; the hinder part of the head, neck, and back, are of a yellowish green; the wing, coverts, and secondaries, are greenish, margined with red; the great quills dusky, the superior coverts and tail of a bright red, the inferior coverts pale ash-colour; the breast is olive-green, and by degrees grows fainter, until it becomes intirely white on the belly; all the under part of the body being striped transversely with broken dusky lines. The legs and feet are pale brown.

The female differs from the male in having a pale yellow bill, the top of the head and neck cinerous; the wings, rump, and back, a yellowish-green, without any tint of red; the quills of the tail brown, margined with dull red.

"Green Gold Finch?"

MOOR BUZZARD.

FALCO ÆRUGINOSUS.	*Lin. Syst. p.* 130. *No.* 29.
LE BUSARD DE MARAIS.	*Bris. orn.* i. *p.* 401. *No.* 29.
LE BUSARD.	*Buff. ois.* i. *p.* 218. *t.* 10.
MOOR BUZZARD.	*Br. Zool.* 1. *No.* 57. *t.* 27.
	Lath. Gen. Syn. i. *p.* 53. 34.

THIS subject is found in the Transbaltic countries, as far north as Sond-mor; it is met with all the year in Sweden, and is common in the south of Russia, but is not to be found in Siberia.

Though smaller than the common Buzzard, it is more vigorous and courageous; and is so rapid and steady in its flight, that it procures a more certain and plentiful subsistence, making dreadful havock among rabbits, young wild ducks, and even fish, and when other prey fails, it will feed on toads, frogs, and aquatic insects.

It builds on the ground in low bushes, or hillocks, covered with thick herbage; its nest being composed of dried sticks, intermixed with sedge or decayed leaves, laying three or four eggs of a bluish white, sprinkled with dusky spots; and though more prolific than the common Buzzard, it is more rare, or at least not so frequently met with.

It does not frequent mountainous situations, neither does it perch on high trees, but rests on the ground, in bushes, hedges, and among rushes, near pools, marshes, or rivers which abound with fish.

The colours of its plumage, and the great length and slenderness of its legs, distinguishes this species from all other of the Hawk tribe.

SPOTTED TANAGER.

Tanagra punctata.	*Lin. Syst.* i. *p.* 316. *No.* 21.
Le Tangara verd piqueté des Indes.	*Bris. orn.* iii. *p.* 19. *No.* 11.
Le Syacou.	*Buff. ois.* iv. *p.* 288.
Spotted green Titmouse.	*Edw.* *p.* 262.
Spotted Tanager.	*Lath. Gen. Syn.* ii. *p.* 228. 20.

This bird is given as large as life; it is described by Buffon under the character of the Syacu Tanagre, being a contraction of the Brazilian appellation *Sayacou*; and by Edwards, as the Spotted Green Titmouse.

The upper parts are green, marked with dusky spots, the rump wholly green, the chin and throat white, the breast the same, inclining to a yellowish tint, the quills and tail brown margined with green, the bill and legs brown.

Spotted Tanager.

CROSS BILL. MALE.

LOXIA CURVIROSTRA.   *Lin. Syst.* i. *p.* 299, *No.* 1.
LE BEC-CROISE   *Bris. Orn.* iii. *p.* 329. *No.* 1.
SHIELD-APPLE, OR CROSS-BILL.   *Edw. Pl.* 305.
                         Br. Zool. i. 115.
                         Lath. Gen. Syn. ii. *p.* 106.

Is represented the size of life; and is found in all the evergreen forests of Russia, Siberia, in Scandinavia as high as Drontliem, in Sweden, in Poland, in Germany, and Switzerland, and among the Alps and Pyrenees; inhabiting only the cold climates, or mountainous situations in temperate countries. It likewise is met with in the northern latitudes of North America, from Hudson's Bay to Newfoundland, arriving at Severn river the latter end of May, but proceeding in a more northern direction, to breed in the pine forests.

This bird is distinguishable from all others by the singularity of its bill, both mandibles of which are in a curve in opposite angles, crossing each other; the upper mandible being blackish, the under mandible gray. This bill, though hooked upwards and downwards, and bent in opposite directions, has, nevertheless, its particular use and advantages, being formed for the purpose of detaching the scales of the fir cones, and the seeds lodged beneath, which are its natural food, and for splitting and tearing apples to pieces, for the sake of the kernels, which is the only part they delight in.

It also assists the bird in climbing up trees, and in raising itself from the lower to the upper bars of its cage, which it performs with such dexterity, that from its mode of scrambling, and from the beauty of its plumage, it has obtained the name of *German Parrot*.

The colours of its plumage are very apt to vary; that of the male in general being red, inclining to rose-colour, mixed more or less with brown, the under part being considerably paler, becoming almost white at the vent, the wings brown, the tail a little forked, and the legs black.

CROSS BILL. FEMALE.

Is of a dusky green colour, more or less mixed with brown in those places where the male is red. Though it is an undoubted fact that these birds change their colours, or more properly the shades of their colours, altering to different varieties of the same colour, yet both sexes still differ from each other, and are easily distinguished at different times of the year.

This bird fixes on the dark dreary forests of pine and fir, as the place of its breeding; and in the depth of winter their loves commence, as they build as early as the months of January and February, and their young are full grown in March, before other birds begin to lay eggs.

They make their nests in the highest part of the fir tree, fastening it to the branch with the resinous matter which exudes from those trees, smearing them with that substance, so that the melted snow or rain cannot penetrate or injure their young ones.

A pair of these birds were kept in this Menagery; they were perfectly familiar, and had a very feeble note; their food was hemp seed, and occasionally the cones of the pine.

In the spring of 1787 a great number of these birds, male and female, were discovered among the fir woods adjoining to the seat of John Eliott, Esq. at Binfield, near Windsor, Berks. Several of them were shot while in the act of feeding voraciously on the cones. They had never been observed at any former period near that spot, nor have they ever since returned.

NICOBAR PIGEON. MALE.

COLUMBA NICOBARICA. *Lin. Syst.* i. *p.* 283. *No.* 27.
LE PIGEON DE NINCOBAR. *Bris. Orn.* i. *p.* 153. *No.* 44.
NICOBAR PIGEON.   *Edw.* 339. *Lath. Gen. Syn.* ii.
          p. 642. 38.

A PAIR of these most beautiful birds were received from the East Indies. They are natives of the Nicobar Islands, which are situated at the entrance of the Gulf of Bengal, and from thence they derive their name.

In size they do not exceed that of the common Pigeon. The bill is dusky; the irides are of an hazel colour; the head, the throat, the thighs, and under coverts of the tail, are of a dark bluish purple; the belly of a dusky brown; the feathers of the neck are long and pointed, like those of the Dunghill Cock, with beautiful and splendid reflections of blue, red, gold, and bronze colour; the back and upper part of the wings are green, changing to copper colour and gold; some of the outer quills, and coverts above them, are of a fine blue; the tail and upper coverts white, the legs reddish.

The female differs from the male in her plumage, which is less glossy; and the pointed feathers on the neck are somewhat shorter.

From the care, attention, and every accommodation which these birds enjoyed in the delightful spot which was appropriated to their residence, there was every reason to hope that they would have bred; but notwithstanding the hen went to nest several times, her eggs were never productive.

MALACCA GROSBEAK. MALE.

LOXICA MALACCA.	*Lin. Syst.* i. *p.* 302. *No.* 16.
LE GROS-BILL DE JAVA.	*Bris. Orn.* iii. *p.* 237. *p.* 15.
LE JACOBIN.	*Buff. Ois.* iii. *p.* 468.
WHOLE-BREASTED INDIAN SPAR-	
ROW.	*Edw.* *p.* 355.
MALACCA GROSBEAK.	*Lath. Gen. Syn.* ii. *p.* 140. 47.

THIS bird measures, from the point of the bill to the end of the tail, four inches and a quarter; the bill is of a bluish ash-colour; the head, the neck, the throat, the middle of the belly, the thighs, and coverts under the tail, are of a rich black; the breast, the belly, and sides, white; the back, the rump, the scapulares, the wings, and tail, are of a bright chesnut; the coverts of the tail of a deeper chesnut, inclining to purple; the legs are brown.

The female differs from the male in having the thighs chesnut, and the colours of her plumage less brilliant.

Malacca Grosbeak . Male.

FRANCOLIN PARTRIDGE. MALE.

Tetrao Francolinus.	*Lin. Syst.* i. *p.* 275. 10.
Le Francolin.	*Bris. Orn.* i. *p.* 245. *pl.* 5. *fig.* 2.
Francolin.	*Edw.* 246.
Francolin Partridge.	*Lath. Gen. Syn.* ii. *p.* 759. 6.

This very beautiful bird is met with in Spain, Italy, and the several islands in the Mediterranean; in Barbary, Egypt, and Asia Minor, and even as far as Bengal in the East Indies.

In size it is rather larger than the common Partridge, which it very much resembles in many respects; yet, upon a close examination, it differs from the Partridge, not only in the colour of its plumage, but in the general shape; and more particularly so in having a spur on each leg, where the Partridge has only a callous tubercle.

Like the Partridge also, it feeds on grain, is very easily reclaimed, and with proper care will breed freely in our menageries. The subject of this Plate was the male; every effort was used to procure a female, but without effect.

STRIATED GROSBEAK.

LOXIA STRIATA.	*Lyn. Syst.* i. *p.* 306. *No.* 37.
LE GROS-BEC DE L'ISLE DE	
BOURBON.	*Bris. Orn.* iii. *p.* 243. *No.* 11.
STRIATED GROSBEAK.	*Lath. Gen. Syn.* ii. *p.* 142. *No.* 49.

THIS bird is something larger, and thicker than the Wren, measuring from the point of the bill to the end of the tail almost four inches; its alar extent is almost six inches.

The upper mandible of the bill is dusky, the under mandible grey, the head and upper part of the bird is brown, the shaft of each feather being rufous; the throat and fore-part of the neck dusky; the scapulares, the coverts of the wings, and tail, are brown, with the shaft of each feather rufous; the breast and belly white; the thighs and coverts under the tail brown; the legs blackish.

Crimson Grosbeak.

ILLINOIS PARROT.

Psittacus pertinax.	*Lyn. Syst.* i. *p.* 142. *No.* 15.
La Perruche Illinoise.	*Bris. Orn.* iv. *p.* 353. *No.* 68.
L'Aputejuba.	*Buf. Ois.* vi. *p.* 269. *No.* 15.
Yellow-faced Parrakeet.	*Edw.* *p.* 234.
Illinois Parrot.	*Lath. Gen. Syn.* i. *p.* 228.
	Arct. Zool. i. *p.* 284.

This bird inhabits the Brazils and Guiana, where it remains the whole year, frequenting the small savannahs, and other open situations; it is also met with at Cayenne, where it is called the *Wood Louse* Parrakeet, because it generally makes its nest, and lodges, near the habitation of that insect; it is likewise common on the banks of the Ohio, and also on the southern shore of Lake Erie.

It is a gregarious bird, being frequently seen in great numbers, of at least five hundred in a flock, which, like rooks, continually place one as an out centinel, to give notice of approaching danger; and when disturbed, they set up an horrible discordant outcry all together.

In length this bird measures nine inches and a half; the bill is of an ash colour; the irides a bright orange; the crown and back of the head, the upper part of the neck, the back, the wings, and tail, are of a fine green, except some of the superior coverts of the wings and quills, which are edged with pale blue; the forehead, cheeks, and throat, are of a bright orange; the fore part of the neck, the breast, belly, thighs, and coverts under the tail, are of a yellowish green; the lower belly yellow; the legs a deep ash-colour; the claws brown.

The subject of this plate was a very lively pleasing bird, exceedingly kind and familiar, very fond of being taken notice of and caressed, and was a very great ornament to this valuable collection.

RED WINGED ORIOLE.

ORIOLUS PHŒNICEUS.	*Lyn. Syst.* i. *p.* 161. *No.* 5.
LE TROUPIALE A' AILES ROUGES.	*Bris. Orn.* ii. *p.* 97. *No.* 12.
LE COMMANDEUR.	*Buf. Ois.* iii. *p.* 214.
RED-WINGED STARLING.	*Cat. Carol. pl.* 13.
RED-WINGED ORIOLE.	*Lath. Gen. Syn.* i. *p.* 428.

THIS bird is a native of America, and peculiar to that quarter of the globe: it is about the size of a starling. They are found in Virginia, Carolina, Mexico, &c. inhabiting the cold as well as warm countries. They appear in Louisiana in the winter only, and at that season are so numerous, that more than three hundred have been taken by once drawing the nets. These nets are spread in some place devoid of grass, at the side of a wood, on which they strew rice and other grain: it often happens that so many are caught that they are obliged to dispatch them on the spot, as it would be impossible to secure so many alive.

They are esteemed the pest of the colonies, making such dreadful havock among the maize and other grain, both when new sown, and when ripe, that they have obtained the name of *Maize Thief*.

They build their nests in bushes and among the reeds, in retired swamps, in a pensile form, leaving it suspended at so judicious a height, and by so wonderous an instinct, that the highest floods never reach to destroy it; the nest is strong, made externally with broad grass, a little plaistered, thickly lined with bent, or withered grass. The eggs are white, thinly and irregularly streaked with black.

The female differs from the male in being smaller, in having a mixture of grey in its plumage, and in the circumstance of the red in her wings being more obscure.

Red Winged Oriole.

White Jerfalcon

WHITE JER-FALCON.

Le Gerfault.   *Bris. Orn.* i. *p.* 370. *No.* 19.
Gyrfalcon.   *Brit. Zool. No.* 47.
White Jerfalcon.   *Lath. Gen. Syn.* i. *p.* 83.

This very elegant species measures very near two feet in length, and is very justly entitled to the first rank among all the birds of falconry, being pre-eminent in courage as well as beauty. Next to the eagle these are the most formidable, the most active, and the most intrepid of all rapacious birds; they are likewise the most costly, and most esteemed for falconry.

They are natives of the inhospitable arctic regions of Europe and Asia, being never found in warm or even temperate countries. They are very frequent in Iceland; are met with at Lapmark and Norway; in Asia on the Urallian and other Siberian mountains, where they brave the coldest climates throughout the year; and support themselves in the open air during the most severe winters, uninjured, in the latitude of Petersburg, when others of the falcon tribe are destroyed by frost; they have been often observed in May about Albany Fort, Hudson's Bay, but retire before winter, feeding on the white grouse, and other birds of the gallinaceous kind, and are sometimes met with (though rarely) in the Orkneys.

In manners and habits of life they very much resemble the Greenland Falcon; they fly at and boldly attack the larger of the feathered race, but their chief objects of prey are the heron, the crane, and the stork, which cannot evade or resist their attacks, and of course become easy victims to their prowess.

COLLARED FINCH.

Le Pinçon a Collier des Indes.   *Bris. Orn. App.* 85. 73.
Le Pinçon a double Collier.   *Buf. Ois.* iv. *p.* 149.
Collared Finch.   *Edw. pl.* 272.
   Lath. Gen. Syn.

This bird is represented on the Plate as large as life, and is called by the Portugueze *Colheerinho.* It is a native of India, and takes the name of Collared Finch from having two half collars, the one before, the other behind.

The bill is very strong and black; the head is also black, except a space round the bill, eyes, and throat, which is white, and passes round the neck, forming a collar; on the lower part of the neck before, is a black bar, which more than half encompasses the neck; the upper parts and tail are of a dark cinereous brown, palest on the rump; the wing coverts are black, margined with a glossy rufous colour, the secondaries the same; the greater quills black; the breast, belly, thighs, and coverts under the tail, white, shaded with faint orange colour; the legs dusky brown.

Collared Finch.

CRESTED PEACOCK. MALE.

Pavo Cristatus.	*Lin. Syst.* i. *p.* 267.
Le Paon.	*Bris. Orn.* i. *p.* 281. *pl.* 27.
Le Paon.	*Buf. Ois.* ii. *p.* 288. *pl.* 10.
Crested Peacock.	*Lath. Gen. Syn.* ii. *p.* 668. 1.

Though the Peacock has been so long naturalized in Europe, it is not a native of this quarter of the globe. The East Indies is the original country of this most beautiful bird, which has been the admiration of all ages, from that of King Solomon to the present time. This opinion is countenanced by Holy Writ, as the Peacock is enumerated among the valuable and rare commodities that were every three years imported from Ophir (probably Sumatra), by Solomon's fleet.

The male is about the size of a middling turkey, measuring from the tip of the bill to the end of the tail three feet eight inches; the female is somewhat less. Above the tail springs an inimitable set of long beautiful feathers, the shafts of each being furnished from its original to its extremity with filaments of a copper colour, and terminates in a flat vane, decorated with what is called the eye, or mirror; this brilliant spot is enamelled with the most enchanting colours, yellow gilded with many shades, green running into blue and bright violet, according as it is viewed in different positions, and the whole receiving additional lustre from the colour of the centre, a rich velvet black; those feathers reach beyond the tail, and are not included in the above measure; the longest of them in the subject now before me measured upwards of four feet: this train or tail, as it it falsely called, can at the pleasure of the bird be expanded quite to a perpendicular upwards, more fully to display their dazzling richness; these are cast every year, either entirely, or in part, about the end of July, and shoot again in the spring, during which interval the bird is dispirited, and seeks retirement.

Nature, in order to form a masterpiece of magnificence, has poured out her treasures with such profusion, and has not only united in the plumage of this bird all that delights the eye, in the soft delicate tints of the finest flowers, all that dazzles in the sparkling lustre of the most precious gems, and all that astonishes in the grand display of the rainbow, but she has selected, mingled, shaded, and blended them, in such a wonderful manner, that they derive from their mixture and their contrast such brilliancy, and such sublime and astonishing effect, that it is beyond the power of the pencil to imitate or describe them.

This bird does not come to its full plumage till the third year; and they are reckoned to live about *twenty-five* by some, while the period of their existence is stated by others to extend to an *hundred years*.

The female lays five or six eggs, the size of those of the turkey; those, if she is left to follow the bent of her own inclination, she deposits in some secret place, to prevent their being broken by the male, which he is apt to do if he finds them. During the whole time of incubation, which is from twenty-seven to thirty days, according to the temperature of the climate, or warmth of the season, she anxiously shuns the male, and is particularly careful to conceal her tract from him when she returns to her nest. The young may be fed with curd, chopped leeks, and barley meal moistened; they are likewise fond of grasshoppers and other insects; in five or six months they will feed on all kinds of grain, like others of the gallinaceous tribe. It has been remarked that the flowers of elder are prejudicial to them, and that the leaf of the nettle is a certain poison to the young ones.

They prefer high trees, tops of houses, and the most elevated places to roost in. Their cry is a perfect contrast to their beauty, being loud and unharmonious.

Gold Crested Warbler.

GOLD-CRESTED WARBLER.

Moticilla regulus.	*Lyn. Syst.* i. *p.* 338. *No.* 48.
Le Pout, ou Souci.	*Bris. Orn.* iii. *p.* 579. *No.* 17.
Le Rottelet.	*Buf. Ois.* v. *p.* 363.
Golden-crowned Wren.	*Edw.* 254.
Gold-crested Warbler.	*Lath. Gen. Syn.* ii. *p.* 508. 145.

THIS very pretty species, in length only three inches and a half, and weighing seventy grains, is the least of all European birds. From having constantly made its nest in a large Portugal laurel in this Menagery, its melodious note, its being quite tame and familiar, it was not only much encouraged, but much valued and admired by the very respectable owner of this most delightful spot, and by his express desire was introduced in the collection of portraits of this Menagery.

END OF VOL. II.

Printed by W. Bulmer and Co.
Russell-court, Cleveland-row,
St. James's.

www.ingramcontent.com/pod-product-compliance
Lightning Source LLC
Chambersburg PA
CBHW021507210326
41599CB00012B/1159